FOREWORD

PETE WATERMAN

Record F  ısiast

I THINK THE LOVE of railways is a very unique thing. My love of trains came right from the beginning. I was born next to a railway line and then you either love them or hate them. I had a grandfather who loved to walk the track and watch the trains go by – that's a strong memory from childhood, that family bond. As a kid from a council house watching trains go by, with people eating food off white tablecloths with flowers and table lamps on, I would think, wow, crikey, that's a different world!

It also makes you feel safe. This thing called the railway just seems to tick. It's comforting and, as a kid you saw it running on time and how everything was just so organised.

You also knew somebody on every street that worked on the railways.

It was a comfort and, particularly when Paul and I were small, the steam engines were massive to us. They roared past, gleaming, noisy and, for little kids, it fascinated me and must have fascinated Paul the same way – it was enthralling.

You went to see grandma on the train, you went to the seaside by train, you got your newspapers, they came by train, the milk came by train. In the modern world everyone has a reason why people do and enjoy certain things, and the word that's bandied around a lot is spectrum. Well, if there's a spectrum we're definitely on it, there's no question about that. Our love of detail is what's made us. It's our love for trains, our love for politics, it's our love for not taking things at face value, but looking at it in far more depth. And a book like this is giving that depth, it gives you a completely different view of the subject from somebody else who is passionate about it.

I never tire of reading books because someone always has a slightly different view than you do. It creates debate, it creates controversy, but it gets you interested. There's this myth that railways are losing their edge, but they're not. Kids are still fascinated by trains. And there's one thing there ain't going to be… an AI trainspotter!

In a world where everything's done automatically there's nothing going to be automatic when it comes to railways. People are still fascinated by railways and, no matter what people say, railways are still the best way to travel. I've been travelling on the railway every week since 1968. For 30 years, five days a week, two journeys a day.

Paul and I have talked about this, we always looked out the window, we were never looking at a screen. So look out of the window. The world changes every single day. It's a wonderful world and on the train you can take the time to really appreciate that.

FOR THE LOVE OF
TRAINS

FOR THE LOVE OF
TRAINS

PAUL ROUTLEDGE

m
B

MIRROR BOOKS

MIRROR BOOKS

1

Published in Great Britain and Ireland in 2025 by
Mirror Books, a Reach PLC business.

Photographic Acknowledgements: Mirrorpix, Alamy

www.mirrorbooks.co.uk
@TheMirrorBooks

Print ISBN 9781917439503
eBook ISBN 9781917439510

Editing and Production: Christine Costello
Cover Design: Chris Collins

Printed and bound in Great Britain by
CPI Group (UK) Ltd, Croydon, CR0 4YY

For the Railway Family, Past and Present

CONTENTS

Foreword 9

Introduction 11

Origins: Taking To The Rails 14

Railway Mania 23

The Great Rail Century – To 1923 33

The Heyday Of The Thirties 41

Rail At War 49

Women And The New Order 58

Smash 68

Dr Beeching's Surgery 78

The End Of Steam 86

Railway Communities 95

Hokey-Cokey Privatisation 105

Trainspotting 113

Trains, Boats, Hotels And Planes 123

The Line That Refused To Die 131

Stations 143

The Royal Train | 151
The Railway In Film And Television | 161
Footplate, Workers | 170
Not Many People Know That! | 179
The Flying Scotsman – A Living Legend | 188
The Railway In Books | 198
Great Railway Journeys | 207
Crime And Cops | 220
The One They Threw Away | 229
The Antique World Of The Railway | 237
Railways Overseas | 247
Great Little Trains | 259
The Great Engine Builders | 269
Go Faster! The Future Of Rail | 283

INTRODUCTION

MORE THAN TWICE as many books have been written about the railway than there are track miles in Britain – 25,000, and counting. So why another one? The only answer is that we all have our own railway, with its own trains, and this is about mine: a love letter from a railway-man's son, born in a railway house, a frequent train traveller, the narrative from a lifelong affair with the industry that changed the world.

Almost my first memory as a child is lying in bed upstairs in 15 Railway Terrace, listening to the familiar clunk-clunk, bang-bang, chuff-chuff of shunting engines and waggons. Work went on 24/7 in the huge marshalling yards of Normanton – in what was then the West Riding of Yorkshire and is now West Yorkshire. It was known as the Crewe of the Coalfields, for the thousands of trainloads dispatched across the country. It was a reassuring sound, a railway lullaby, and I hear it still.

Coal was the commercial midwife of the railway, making fortunes for the mine owners. But Normanton was more than a glorified goods yard. It was the first railway hub, pioneering trains from London to Scotland in the 1840s. In the station's lavish dining rooms, travellers could eat amid Victorian splendour, or make their way over a covered walkway to a hotel for drinks. For more than 100 years, crack expresses like the Thames-Clyde from St Pancras to Glasgow called at this otherwise undistinguished coalmining town.

Not any more. The demise of King Coal was matched by the slow decline of the railway it served. In the aftermath of Dr Beeching's drastic surgery of the system in the 1960s, the ornate buildings were demolished. The express trains were diverted via the county town of Wakefield and the direct link to Sheffield and London was severed when Goose Hill cutting was filled in. All that remains is an unstaffed halt, with local trains to Barnsley and Leeds.

But if it's in the blood, it never goes away. In my class at the local grammar school in the fifties, your father was down the pit or on the railway, and my father Harry was a life-long employee, first of the London, Midland and Scottish and then British Railways. His father, Jack had been a colliery winding engineman, operating steam-driven engines on the surface at pits in the Castlefield area, rather like a stationary railway. I could never escape my inheritance.

Nor would I wish to. The romance of the railway has endured for 200 years since 600 excited, apprehensive, brave souls took their seats in open-topped waggons for the world's first passenger train service from Darlington in England's north-east County Durham, to Stockton, barely ten miles distant.

The bicentenary year of that epic event, commemorated on

29 September 2025, is an excellent time to write "one man and his railway", amid a full year of celebrations, ranging from a replica run to the greatest gathering of locomotives for generations.

The glory days of steam and speed may have passed away, but the lure of the train lives on.

CHAPTER 1

ORIGINS: TAKING
TO THE RAILS

IT WAS A novelty, a spectacle, drawing more than 10,000
spectators on that autumn morning. Little did they realise
that they were witnessing the dawn of a new age. They
scrambled for the best places to watch George Stephenson's "superior locomotive engine" Locomotion No 1 pull
a train of goods waggons and wooden coaches, over-loaded
with passengers, from Darlington to Stockton in County
Durham, at speeds reaching 12 miles an hour, unheard of
in the era of the coach-and-four. This date, 27 September
1825, marked the beginning of rail travel and it changed the
world.

Private lines worked by horses or stationary engines
had been in use within collieries for many years, but this
was the first time a steam locomotive on wheels had been
seen in public. It was a sensation. Some were disappointed
with the new phenomenon, expecting to see an "automati-

cal semblance" of a horse stalking on four legs. When the engine driver stopped to take on water and let off steam, the crowds fled in fear of an explosion – or perhaps dreading something more "uncanny" from this strange beast come into their midst.

The practical, hard-headed men in charge of the enterprise had no such fanciful notions. Their objective was to move coal from pit to market at the cheapest, and most profitable, price. Passengers were an afterthought, not even specified in the original prospectus, but they swiftly realised that a new, and highly-commercial opportunity was opening up: truly, the first age of the train.

It's hard to emphasise just how revolutionary the railway was, though JS Jeans, an official historian of the Stockton and Darlington Railway, gave it a go: "Thus began the greatest innovation of the 19th century. Galileo, when in the face of the Inquisition that threatened his life he proclaimed 'it moves still', was not considered more rash and iconoclastic in his day and generation than were the apostles of the railway system in theirs. They were, indeed, the true iconoclasts. They destroyed pernicious errors and gross delusions."

It may be said to have begun, not on rails, but on water. Canals were the great transport solution of the 18th century, with many hundreds of miles constructed by "navvies" (navigators) by pick and shovel across the country. The building of a waterway from the mines of west Durham was proposed, but the new men of commerce had a better idea, and at a "highly respectable" meeting in Darlington Town

Hall on 13 November 1818, they prevailed. On that fateful day, the canal age ended.

A committee of management was appointed, which drew up a prospectus for "a rail or tramway throughout the entire line between Stockton and the collieries". An application to Parliament sought approval for a Bill to authorise the scheme. The cost was put at £124,000, in shares of £100 each. They evidently had no trouble raising the money and surveying began.

The work soon encountered strong opposition from landowners, in particular Lord Darlington of Raby Castle, who refused permission for the route, on the grounds that a railway would be "harsh, oppressive and injurious of the country through which it is intended to pass." The hunting fraternity was also angry that the line would go through the Duke of Cleveland's fox covers, which were of course vastly more important than a railway for the hoi polloi. The promoters pulled all the political strings at their disposal, but their Bill fell at its second reading. Had they lost heart, others would no doubt have taken up the torch, but they persevered, and a second survey led to another Bill.

This one was much shorter, and the management committee reported "with extreme satisfaction" in a manifesto of November 1820, a favourable change in public opinion, with very few dissenting voices now heard. The death of the King George IV delayed the project yet again, but the Bill finally received the Royal Assent in April 1821.

Edward Pease, a local Quaker textile manufacturer and arguably the key promoter, secured the services of George

Stephenson, a Newcastle man-in-a-hurry currently laying lines at nearby Hetton colliery, as the line's engineer. A second parliamentary Act, passed in April 1823, gave powers to make and use "locomotives or movable engines for the transport of goods, articles and merchandise and other articles....and the conveyance of passengers." This virtually-throwaway line was the first time the promoters had even hinted that they were intent on building a passenger railway.

George Stephenson lost no time in laying out the line, assisted by his son Robert, by one account "a slight, spare, bronzed boy" taken from his work as a pit-viewer and whose knowledge was criticised as no better than any other very intelligent but inexperienced lad. He later achieved fame as one of the greatest engineers of the 19th century.

The Stockton & Darlington's pioneer engines, Locomotion No 1 and Hope, costing £500 each, were built at Forth Street Engine Works, Newcastle, under the direction of Stephenson pere. The 18-ton engine and coal tender was designed to draw on a level 48 tons gross load at five miles an hour. Unimpressive, in today's terms, but then a phenomenal surge in productivity – and profit.

Construction continued swiftly despite opposition from a few stubborn landowners, whose objections had to be overruled by recourse to a jury's valuation, "soon terminated in a manner favourable to the company". Some contractors had to be paid bonuses to keep up the pace, and on 9 September 1825 the promoters admitted to shareholders that expenditure had far exceeded the engineer's estimates. But nothing would stop them now, and the premier run of the world's first passenger train was fixed for a little over a fortnight later.

Even now, dissenting voices carped that it would never catch on. Lord Eldon, claimed that railroads signalled that "Englishmen who were wont to be sober are grown mad", while one participant in the scheme dismissed talk of speeds of 12, 16, 18 or 20 miles an hour as "ridiculous nonsense".

The company lost no opportunity to publicise the historic event. Local newspapers carried an announcement that, after collecting coal waggons on steam-worked inclined planes from the colliery nine miles west of the town, the train would leave Darlington at 9.00 precisely and proceed by way of Yarm to Stockton on Tees, where it was calculated to arrive about one o'clock. An elegant dinner would be provided at the Town Hall, for the neighbouring nobility and gentry who had taken an interest in this "very important undertaking". A superior locomotive would haul convenient carriages for the conveyance of proprietors and strangers. Gentlemen (not riff-raff) might apply for permission to ride.

A company handbill was more enlightening about the mystery train. The starting point was actually nine miles west of Darlington, at Brusselton Tower, atop two inclined planes powered by stationary engines to draw up and then down loaded coal waggons from Witton Colliery. Spectators were bidden to assemble there at nine o'clock to see the formation: the company's locomotive engine, the tender with water and coals, six waggons, laden with coals merchandise etc, the management committee, and other proprietors in the coach belonging to the company, six waggons with seats reserved for strangers and 14 waggons for the conveyance of workmen and others: The Whole of the above to proceed to Stockton.

It was more of a cavalcade, a victory parade, even, than a humble train. Six waggons laden with coals would leave the train at Darlington. More waggons "for workmen and others" would be drawn by horses. The company's workmen would leave the procession at Darlington at one o'clock, and dine there. Tickets would entitle them to eat at specified Houses of Entertainment. Talk about class distinction!

One can easily imagine the boisterous, bibulous scenes in the town's "specified" drinking establishments. I saw them myself, as a young trainee reporter on the *Northern Despatch*, 140 years later. "Darlo" knows how to enjoy itself.

The formal opening notice continued: "The PROPRIE-TORS. And such of the NOBILITY and GENTRY as may honour them with their company will dine precisely at THREE o'clock in the TOWN HALL, STOCKTON. Those wishing to return to Darlington would be taken back at 7pm. Precisely."

A four-hour lunch, for the toffs, with no caution as to their gormandising. It was a different matter for the men who built the line. The bill's penultimate paragraph warned: "The COMPANY take this Opportunity of enjoining on all their WORK-PEOPLE that attention to Sobriety and Decorum which they have hitherto had the Pleasure of Observing."

Finally, the management committee gave notice that all persons who shall ride upon, or on the sides of, the RAILWAY, on horseback, will incur the penalties imposed by Acts of Parliament. Thus began the prohibition of setting foot, or on horseback, on the permanent way, as the running track became known, immortalised for two centuries by large cast-iron 'No Trespassing' signs bearing the name of the railway company, which disappeared only a few years ago and are now much sought-after mementoes of a forgotten era.

As recorded by the company chairman, the final make-up of the train comprised: Locomotion and tender, five waggons laden with coal, one with flour, one with engineers, the management committee and friends in "the long coach", six waggons with strangers, 14 waggons with workmen and others and six waggons laden with coals. Even by contemporary modest standards of rolling-stock construction, this was a mighty show of triumph and prosperity.

Flags with inscriptions decorated four of the waggons, one reading, "Periculum privatum utilitas public" – The private danger is the public benefit – and another, "May the Stockton and Darlington Railway give public satisfaction and reward the liberal promoters."

An official history of the occasion, published in 1875 on the 50th anniversary, records, "The universal cheers, the faces of many, the vacant stares of astonishment of others, and the alarm depicted in the countenances of some, gave variety to the picture." The train, weighing at least 80 tons, moved initially at a speed of ten to 12 miles an hour, but on the downhill final stretch accelerated to 15 or 16 miles an hour. Some poor souls must have felt they were going to hell in a hand-cart. Fewer than 300 tickets had been given to applicants, but almost double that number got into or clung on to the waggons.

The journey from Darlington to Stockton took three hours and seven minutes, with stops, including a long one at Yarm to drop off some coal waggons: business, already? Spectators lined the 12-mile route, gaping from fields, lanes, bridges, any vantage point to see the historic event. As it neared Stockton, the cavalcade was joined by a motley procession of horses, carriages and people on foot. The scene resembled a coronation.

Official historian JS Jeans, wrote of the first public train, "As a precaution against possible accident, and, perhaps also, to give a more imposing effect to the procession, the company engaged men to ride on horseback in front of the engine. These heralds held flags in their hands, and gave notice to all who it concerned that the locomotive was approaching."

On arrival at the company's wharf in Stockton, a 21 gun salute was fired and, the band striking up 'God Save the King', was greeted by a "three times three stentorian cheers." Oh, to have to have seen that day.

Contrary to the nay-sayers who scorned that "it would never catch on", the railway was an immediate commercial success. The price of coal to the public was reduced by a third, and the company directors boasted of a huge demand, including a single order from London for 100,000 tons. Shares now commanded a £40 premium, and there were "plenty of buyers, but no sellers".

After the initial excitement, however, passenger traffic was slow to build. There had never been much demand for the coach between the two towns and people were not accustomed to frequent travel.

Indeed, the company had no license to carry passengers when it opened and didn't apply for one until the following month. Services began on 10 October 1825, with a new coach, Experiment, every day except Sunday.

There were unforeseen problems. The enginemen had to contend with unofficial passengers: people who ran alongside and jumped on to the waggon for a free ride. This practice becoming "intolerably common", the company put up notices threatening penalties, but it was also thought that enginemen might be conniving, and profiting. They, too, received a warning.

By 1832, passenger figures had risen to a modest average of 520 a week, paying one and a half pence inside and 1d outside per passenger mile, with a maximum of six inside and 20 outside.

At least it was safe. In its first seven years of existence, the S&D carried passengers to a total length of 300,000 miles without injury to life or limb. To the passengers, that is. For the men on the engine, this new occupation was a more risky business, as we shall see.

CHAPTER 2

RAILWAY MANIA

THE EXCITEMENT THAT greeted the world's first public railway rapidly developed into a fever. What became known as "railway mania" gripped the nation. Those with money to invest could not get their hands on shares for new routes fast enough.

Big spenders sank three billion pounds into new lines reaching into all parts of the country, some duplicating each other, to big cities and rural communities that couldn't even sustain a stagecoach. Fortunes were made – and lost. Corruption and fraud ran through the industry faster than Locomotion No 1, as competition from rival promoters reached lunatic levels. Powerful businessmen like George Hudson, "the Railway King" dominated trade.

By 1875, 50 years after the far-sighted Quaker pioneers of Darlington, like Edward Pease, had made history, it was estimated that the railway now covered 160,000 miles, built at an average cost of £20,000 a mile, totalling £3.2billion pounds.

From a humble stable of two locomotives, numbers now exceeded 50,000, equivalent to the power of ten million horses, still an understood comparison in Pease's days. Employing fewer than "three hundred hands" in 1825, the industry now found work for 300,000 in Great Britain alone.

The seeds of this remarkable expansion had been sown at the S&D company's celebratory dinner in Stockton Town Hall on 27 September 1825. Around the lavish table, seated alongside the promoters of the line and local nobility, were chairmen of the newly-projected Liverpool and Manchester Railway and of the Liverpool and Birmingham Railway.

They were impressed by what they had seen that day and resolved to do likewise. George Stephenson, architect of the route, had already offered his services to the Liverpool and Manchester, which was to become an even more controversial undertaking than his first foray into public railways.

It was a battle on three fronts. First, he had to contend with the greed and ignorance of landowners, who opposed the world's first genuine inter-city route on specious grounds that it would make their cows run dry and frighten the rural populace, and the understandable desire to grasp as much compensation as possible from the promoters.

The second battle was just how to get the line across territory that nature made hostile, particularly Chat Moss, a boggy morass that swallowed anything that moved. It proved impossible to drain completely, so millions of tons of spoil were dropped into the sodden void in one place for three months without finding the bottom.

But his third front was scientific. Could these new-fangled

steam engines be strong and reliable enough to provide a consistent service?

To settle the question, the company held the famous Rainhill Trials, offering a £500 prize to the engineer who could make a locomotive weighing no more than six tons pull a load of ten tons at ten miles an hour. That test excluded George Stephenson's Locomotion No 1, which weighed eight tons.

Three entrants took part in the competition: Novelty, which was really only a steam car on wheels that frequently broke down; Timothy Hardwick's Sans Pareil suffered a leak that had to be stuffed with porridge. And the winner was the legendary Rocket, built by George's son Robert, to a revolutionary design incorporating a multi-tube boiler for the first time, vastly increasing the steam pressure.

Rocket proved capable of speeds of up to 29mph with the requisite load, and astonished onlookers by storming up Sutton Incline, a feat hitherto regarded as impossible.

The Liverpool and Manchester opened for business on 13 September 1830, amid celebrations marred by the first passenger railway fatality – and not just anybody, but Liverpool MP William Huskisson, former President of the Board of Trade, whose unhappy end we notice later.

The service prospered mightily, boosting company shares and creating a new class of business commuters, who could travel between these two great cities in two hours on a daily basis. Raw materials could be transported from Liverpool docks to the great factories of Manchester more cheaply and at much greater speed than by canal. The coach companies went out of business, and mail was carried by train for the first time. The Railway Age had truly arrived.

Predictably enough, it was swiftly followed by that Railway Mania of the 1840s, when the runaway success of the Liverpool and Manchester prompted a plethora of get-rich-quick schemes that yielded a vast expansion of the system from London to Liverpool, Manchester, Bristol, Glasgow, Edinburgh, Cardiff, Norwich, Leeds and a host of other towns and cities. But it also triggered a stock market bubble and a parliamentary frenzy. In 1846, no fewer than 263 Acts of Parliament creating railway companies were passed. A third of these never materialised, or were swallowed up by rival companies before a mile of track was laid. Some turned out to be sham outfits, gulling greedy investors into parting with their savings.

Attempting to hold back all this progress, prominent objections still came, many using arguments of which today's NIMBYs would be proud. MP Charles de Laet Waldo Sibthorp, popularly known as Colonel Sibthorp and a living caricature of a Tory MP, said, "Next to a civil war, railways are the greatest curse to the country."

Poet William Wordsworth wrote in an 1844 letter to William Gladstone, then President of the Board of Trade, "We are in this neighbourhood all in consternation, that is every man of taste and feeling, at the stir which is made for carrying a branch Railway from Kendal to the head of Windermere.

"The project, if carried into effect, will destroy the staple of the Country which is its beauty, and, on the Lord's Day particularly, will prove subversive of its quiet, and be highly injurious to its morals."

The line opened in 1847, encouraging development of a lakeside resort attracting up to 8,000 trippers on Bank Holidays in the 1880s. The effect on morals is unrecorded.

Art historian John Ruskin was equally appalled by the new railway in the rocky valley between Bakewell and Buxton in Derbyshire, saying, "The valley is gone, and now, every fool in Buxton can be at Bakewell in half an hour, and every fool in Bakewell at Buxton: which you think a lucrative process of exchange, you Fools everywhere." The line closed in the 1960s Beeching era.

The railway boom was always going to end in tears. Investors could put down only 10% to buy shares, with the company able to call in the full amount when they pleased. Anyone could start a scheme, however fanciful, without owning the land or showing any expertise in railway construction. Promoted as risk-free investment, including by newspapers read by an increasingly-literate middle class, speculators took advantage of a free-for-all allowed by a permissive government after the repeal of the Bubble Act, designed to prevent a repetition of the 1729 South Sea Bubble, in 1825.

Some MPs had their hand in the till, investing heavily in the schemes they rushed through Parliament. A laissez-faire state allowed the new breed of capitalists a free hand, laying down no regulations except for the universal four-foot-eight-and-a half inch gauge standardised by George Stephenson. All complied, save the Great Western, whose incurably romantic pioneer Isambard Kingdom Brunel insisted on a monster seven-foot gauge, which remained in remote parts of the network until 1892.

The disaster that was waiting to happen duly took place in the late 1840s. Bank of England interest rates rose, banks put money into bonds, investment in railways slowed down to a pace lower than Locomotion No. 1 and then halted in its tracks. Bills for new companies fell to zero, and larger firms like the Midland and the GWR began buying up hopeless ventures to consolidate their operations.

The boom was over, but it had done its work. Lines totalling 6,220 miles (around half of present-day track mileage) were built on the back of legislation passed between 1844 and 1846. Investors came back in subsequent years, with more realistic ambitions, and by the end of the 1860s, the national network was essentially complete, with only branch lines and ventures like the West Highland awaiting construction.

The close of the 19th century may have been relatively quiet, in terms of new lines, but with rival routes offering direct travel from London to Edinburgh, and the introduction of more powerful locomotives capable of long stretches at high speed, it was inevitable that cut-throat competition would emerge. It began in earnest in June 1888, when the companies operating the east coast main line from Kings Cross to Edinburgh Waverley and the west coast route from Euston to Edinburgh Princes Street, began to cut journey times.

In June that year, the west coast operators fired the starting pistol, retiming their Day Scotch Express to arrive an hour earlier, at 19.00.

Their rivals retaliated on 1 July with the Special Scotch Express arriving at 18.30, partly by reducing the half-hour refreshment stop at York by ten minutes, only for the competition to follow suit a month later. But on 1 August, in a fit of one-upmanship, the east coast operators slashed their arrival time to 18.00. The west coast companies came back with equally-fast timings, though with two separate trains, one with fewer carriages dashing from London to Crewe non-stop.

The race was on! It went into the final straight on Bank Holiday 6 August, amid a frenzy of media interest. Who would take the palm?

Bookies took bets, and the public took their places for a classic show of railway theatre. Reporters raced the mile between Princes Street and Waverley to describe the scene. The East Coast cut out the stop at Berwick and further slashed its timing, with a booked arrival at 17.45, surely a record?

The West Coast abandoned all pretence of timetable and beat their rival before the record was even set. Mania was the correct word for this drama on the rails and on 14 August the companies met to agree to an armistice. The race was over – for now.

Roger Fulford, writing in *The Listener*, said, "The East Coast route won the race, and it was a superb performance. Honour satisfied, the railway companies agreed to a truce - seven and three quarter hours for the east, eight hours for the west."

The companies sank back on their laurels, exhausted but pleased with the publicity gained for their services. But rivalry broke out again on 15 July 1895, weeks ahead of the opening of the grouse season.

This was a critical factor in the equation. In late summer, the sporting gentry headed for the moors and hills of Scotland for game shooting. These wealthy patrons travelled with a whole retinue of families and dogs, plus mountains of luggage full of weaponry and victuals: the "grouse traffic" in railway terminology.

The second Race to the North went further than the first, to Aberdeen, not far from Queen Victoria's summer residence at Balmoral. It began in virtual secrecy, uncovered by a clergyman at Euston station who saw new posters advertising that the West Coast 20.00 service to the Granite City would now arrive at 07.00, an hour ahead of the timetable. Our man of the cloth scurried down the road to Kings Cross, where he found East Coast managers preparing a response to this coup de guerre, and on 21 July they announced that their express to Aberdeen would now reach its destination at 06.45.

The race was on, yet again. The West Coast had an advantage, in that only two companies were involved, the London North Western and the Caledonian, each fully engaged in the underhand competition. They had to change engines only once, at Carlisle, whereas the East Coast had three companies and two changes: the Great Northern to York, the North Eastern to Edinburgh and the North British thence to Aberdeen. The situation was even more complicated by the fact that the two rival routes converged north of the Firth of Forth at an obscure spot, Kinnaber Junction, just north of Montrose. Whoever got there first, won the race – for that day. Competition hotted up as "the 12th" neared, with the West Coast surreptitiously speeded up by 25 minutes to arrive at 06.35. The East Coast retaliated with 06.25, and the West hit back with a further five minute reduction, and on 30 July set a new record, rolling into Aberdeen at 05.59 – less than six hours for a 550-mile journey.

The climax to this game of railway madness appeared to be close in mid-August, when the rival trains met in timetable combat at Kinnaber Junction. On the night of the 15th, the signalman was offered the West Coast one minute before the East – and on the following night both bells rang simultaneously, leaving the poor signalman, an employee of West Coast company Caledonian, Solomon's choice of which to allow through first. He gave the road to the East, later described as "a fine example of the sporting spirit shown throughout", though it must have been a nerve-wracking experience.

And the race wasn't yet run. The rivals continued cutting running times, abandoning all pretence of a timetable. On 21 August, the East Coast reached the tape at 4.40, beating the West by almost 15 minutes. The next night, the West ran a stripped-back special that took only eight hours and 32 minutes, at an overall speed of more than 60 miles an hour.

The race was over, with honour satisfied all round – except for the poor passenger who found himself in Aberdeen station at the crack of dawn with nowhere to get breakfast after a breakneck nocturnal dash that upset his digestion. One disgusted wealthy customer wrote to his newspaper, "I had to travel in a racing train and I reached Aberdeen in ten hours. The oscillation was so great that I felt sick. Two of my servants were sick. A friend of mine only saved himself from sickness by a dose of brandy."

It had been railway theatre at its best, with large crowds gathering at Euston and Kings Cross every night to wish the driver and his fireman a noisy bon voyage. The footplatemen who achieved these remarkable feats got no publicity and the race itself never actually took place, according to the published timetables that went unchanged throughout the weeks of warfare. Indeed, the LNWR chairman told a board meeting, "There is no such thing as a race. But our company will not be the last in it."

With Victoria's reign and the century, drawing to a close, the railway was firmly established as the nation's primary means of transporting people and goods. It was almost a state within a state, answerable only to its shareholders and, only when compelled, to Parliament. Its primacy, and profitability, was soon to come under challenge.

CHAPTER 3

THE GREAT RAIL CENTURY – TO 1923

WHAT WAS LATER celebrated as the so-called Golden Age of Travel, when it was possible to travel by train from one end of Europe to the other in some comfort and without a passport, ended with the assassination of Archduke Franz Ferdinand of Austria in Sarajevo (incidentally reached by the railway in November 1882) on 28 June 1914.

The Great War started afterwards, a catastrophe for mankind, but arguably the financial saviour of the train in Britain. Railway companies had begun losing money in the 1870s and ended the 19th century amid calls for nationalisation. These were not new.

As early as 1844, with the "railway mania" surging through the city, there were calls, duly ignored, for the government to take over. By 1890, author F Keddell, writing in his book *The Nationalisation of Our Railway System: Its Justices and Advantages*, could condemn the laissez-faire system in the strongest terms, "The only ground for maintaining the monopoly would be the proof that the railway companies have made a fair and proper use of their great powers, and have conduced to the prosperity of the people. But the exact contrary is the case."

The war changed everything, and quickly. Hostilities with Germany opened on 28 July 1914, and on 4 August the British government took over the entire network under the control of a Railway Executive Committee. Managers from the big operating companies – the people who knew how to run the trains – took part in what was an embryo of nationalisation. The railway assumed a paramount importance, militarily and industrially.

The Committee put into operation a secret plan to embark the British Expeditionary Force to France. By the end of the month, 670 special trains had ferried 118,000 men, with horses, guns and other impedimenta of war to the docks, a feat requiring one train every 12 minutes for 16 hours a day.

Then the full impact of total war fell on the shoulders of managers and men whose ranks were considerably depleted by volunteers for the army. The railway wasn't a "reserved occupation" in this conflict.

Demand on the network increased exponentially, with troops trains, ammunition trains and even secret trains like the Jellicoe Specials, deployed to take steam coal from the Welsh mines 700 miles to the northern tip of Scotland for the Home Fleet based at Scapa Flow.

At the outbreak of war, about 13,000 women worked on the railway, chiefly in what was then called "women's work" like cleaning, washing and waitressing. As the war went on, more and more stepped in to fill the gaps left by the many thousands of railwaymen who enlisted. In all, an estimated 1.6 million women took over jobs on the system, as engineers, in sheds, on stations and in offices, doing work that had hitherto been exclusively for men. Many remained in their jobs after the war.

Railway workshops were given over to war production. There was a huge demand for locomotives. Some of the fleet were ancient single-wheelers, suitable for racing to Scotland with a couple of carriages but not much good for long trains of coal, steel and military material like tanks. The authorities decided that a new design was necessary and adopted chief mechanical engineer of the Great Central Railway John G Robinson's no-nonsense, rugged 2-8-0 freight engine as the workhorse of the war. A total of 521 were built, with many serving in the overseas theatre. Most survived into nationalisation in 1948, and one, 3601, is preserved in the National Railway Museum in York.

Congestion on the system was a huge logistical problem and was a factor in the infamous Quintinshill rail disaster, the country's worst-ever in the early morning of 22 May 1915. Five trains were involved, including a troop train crammed with soldiers of the Royal Scots heading for Gallipoli. The old, gas-lit wooden carriages caught fire, and 227 people died: 215 soldiers, nine other passengers and three railwaymen. Two signalmen blamed for lax behaviour that caused the disaster were convicted of manslaughter and jailed. They were re-employed after their release in 1916, but not as signalmen. A BBC documentary, aired in

2015, blamed the company for over-loading the track with peacetime services as well as troop specials to maintain profits. As casualties mounted, ambulance trains became a regular feature. Transformed into mobile hospitals, they treated six million wounded servicemen, a large number of them abroad.

After the war was over, the railway counted their dead: 20,000 lost on active service. And the politicians did their sums about the future of the system. It was clear that the ramshackle collection of competing companies was unfit for purpose and had been for years. Most were losing money, and something had to be done. Within six months, Transport Minister Eric Geddes, a former LNER executive, rejected nationalisation in favour of amalgamation of the failing firms into a handful of privately-owned regional monopolies. Naturally, the companies and Tory MPs bitterly opposed the idea, but it passed into law in 1921.

Two years later, 1923 ushered in "the Big Four": the London and North Eastern Railway comprising the main line from Kings Cross to Scotland and most of the east of the country; the London Midland and Scottish, swallowing the west coast, the West Midlands, north-west England, most of Yorkshire and much of Wales; the Great Western was essentially the same as the old GWR; while the Southern served most of southern England. Only a few minor routes like the Somerset and Dorset – the "Slow and Dirty" – remained outside the new set-up. The new big boys inherited 19,585 miles of track and rationalisation was inevitable and sensible, but it failed to save the railway from the march of history.

The era of the lorry and the motor car had arrived. Road travel became cheaper, the transport of goods easier. The Big Four never made a decent return on capital invested and the LNER never once made a profit (as indeed it doesn't now, being subsidised by the taxpayer after three failed attempts at privatisation).

A very different threat came from within: the railwaymen themselves, engaged in the class struggle that had gone quiet during the Great War, but now resurfaced with renewed strength. Trade unions bringing workers together from all over the system had existed almost since the dawn of the railway age. Engine drivers on the North Eastern formed a society in 1865. It was broken by the company after an ill-advised and ill-funded strike, but a more coherent organisation of depot branches in 1880 saw the emergence of the Associated Society of Locomotive Engineers and Firemen, still going strong today. The rival National Union of Railwaymen was founded in 1913 by the amalgamation of three smaller unions dating from 1872. The picture was completed in 1897 with the formation of the Railway Clerks' Union.

These three unions dominated an industry of hundreds of thousands of men, with close links to the mining unions. Railwaymen and miners often lived cheek by jowl in communities dedicated to the work. Their unions came together in a Triple Alliance in 1914, pledged to support each other in times of trouble. It was a recipe for collaboration in a titanic struggle against the state, which duly came with the General Strike of 1926.

There had been portents of what was to come. In 1901, the

Taff Vale Railway won a court action for lost earnings due to a strike, making unions liable to potential bankruptcy. This blow led, within a few years, to the formation of the Labour Party, dedicated to seeking parliamentary representation for workers and the reversal of the Taff Vale ruling. In 1911, the unions co-ordinated their actions in a two-day strike that forced the Liberal government to recognise them.

After war's end, the unions pressed for nationalisation of the railway, but the companies were passed back into private hands. Their demands for higher wages and an eight-hour day prompted a nine-day stoppage in 1919, winning most of their claims.

Meanwhile conditions in the mines were deteriorating rapidly and when they were decontrolled in 1921, pay cuts were imposed on the pitmen – but the rail unions refused to come to their aid, on a date known in labour movement history as "Black Friday".

Honour of a sort was restored in the General Strike five years later, but amid lasting acrimony the politically-ambitious NUR leader JH Thomas worked hard behind the scenes to end the national stoppage that shook the nation. The railwaymen loyally came out in their hundreds of thousands, but the transport walkout was called off after nine days, leaving the miners to fight on for many months before being starved back to work.

The General Strike had its light and dark moments on the railways. Undergraduates and former army officers volunteered to drive trains, unaware of the skilled nature of the job. They gave the newspapers photo opportunities for amateurs on the footplate, but they also killed four people in accidents at Edinburgh and Bishop's Stortford. A volunteer driver in Hull was severely injured when his train ran into parked waggons.

But there was also violence against trains that did run. The "Flying Scotsman" hauled by A3 Pacific 2565 Merry Hampton with 281 passengers on board was derailed at Cramlington, Northumberland, on 10 May, the seventh day of the strike as a result of sabotage and seven people were taken to hospital with minor injuries. Striking miners had removed a section of the track, hoping to prevent transit of a coal train.

The volunteer driver of the express had been alerted to possible trouble ahead and slowed down but his engine and several coaches came off the track. The community closed ranks and no one would grass up the culprits. However, eight miners – Arthur Wilson, William Muckle, William Baker, Thomas Roberts, James Ellison, Robert Harbottle, William Stephenson and Oliver Sanderson, all in their 20s – were subsequently jailed for a total of 48 years for their part in the near-tragedy, serving their sentences in far-off Maidstone prison, the last three "train wreckers" freed by the new Labour government on condition that no political demonstrations would be held in the capital. They were feted on return to Newcastle station.

The real damage to the railway, apart from the revenge meted out by company managers, came from the acceleration of the transition to the roads of passengers and goods. As one car manufacturer put it, "Motoring has once and for all knocked the possibility of a serious transport strike on the head. With half a million capable motor drivers in the country, it is an anachronism."

On the strict point, he was wrong. Even with 35 million drivers on the UK's congested roads, there are still rail strikes and they are still crippling, but the rapid shift of traffic from rail to road in the "Roaring Twenties" hit the Big Four hard. No new lines were being built but taxpayer-built roads and cheap war-surplus vans and lorries ate into their profits. The economy of the railway was also undermined by the government's refusal to free the companies from the "common carrier" duty, imposed in the 19th century when they had a virtual monopoly, to take any cargo at an agreed national charge, usually less than the cost of providing it. Road hauliers enjoyed the freedom to charge what they liked and what they liked was a cheaper service, and door to door. The railway companies retaliated by going into the road haulage business themselves, switching from horse-drawn deliveries from the goods yard to vehicles such as the Scammell "mechanical horse".

The twenties drew to a close amid great foreboding. In late October 1929, the Wall Street Stock-Market Crash in America triggered the most severe economic depression ever experienced in the Western world. After an election in which women under 30 took part for the first time, Ramsay Macdonald formed his second minority Labour administration – with renegade rail union leader JH Thomas, hated for his subversive role in the General Strike, as the Lord Privy Seal.

CHAPTER 4

THE HEYDAY OF THE THIRTIES

THE THIRTIES WERE the decade of defiance, a time of speed, style and romance for the railway. New titled trains, new streamlined locomotives that took on the look of aeroplanes. It was the spirit of the Victorian Race to the North, but with bigger loads, on scheduled expresses, not stripped-back racing stock.

They had to do something. The roaring twenties ended in economic depression, and the challenge from the road grew exponentially. Yet the thirties were in some respects the heyday of the railway. The great locomotive designers were given their head and produced the finest examples the world has ever seen. Press and broadcasting publicity and marketing became key weapons in the fight for survival.

Companies formed their own road haulage operations, and even went into the airline business. In the inter-war years, the railway was at its most imaginative since the Race to Aberdeen.

On the main lines, automatic electric lights replaced the old semaphore signals, enabling faster running and cutting down on the boxes needed to control traffic. The great locomotive engineers of the day – Nigel Gresley on the LNER, William Stanier on the LMS – competed to produce the most powerful, long-distance trains. The race to Scotland was back on, with mighty Pacifics, not single-wheelers.

This was truly the heyday of steam, often called The Golden Age of the Big Four. It was also the dawn of an era of new technology, driven by advances in Germany and the USA. Dr Diesel's invention was proving its worth on the track with reliable, fast traction. Third-rail electrification made great strides, particularly on the commuter routes of the Southern. Diesel railcars appeared on the GWR, but mainly branch lines. The writing was on the wall for steam, but a railway on an island built on coal stuck to tradition for another half a century, giving ground to the "infernal combustion" and electric power only slowly.

The 1921 Railway Act that set up the Big Four after The Great War unleashed a new sense of purpose. The network now covered 19,600 miles, reaching into every corner of the country. This was the era of opportunity. William Stanier arrived on the LMS from the GWR, bringing Swindon know-how to Crewe. He designed and built a range of passenger, mixed and goods engines that transformed the locomotive fleet, beginning with the taper-boilered two-cylinder 2-6-0 of 1933 and the Princess Royal Pacifics for the Euston-Glasgow trains. His three-cylinder 4-6-0 Silver Jubilee class, ideal for the St Pancras to Glasgow Thames-

Clyde express, followed a year later and the mighty, stream-lined Princess Coronation class in 1937.

Arguably his major contribution were the two workhorses of the system – "the engines that won the war." They were the go-anywhere, pull-anything Black Five, first outshopped in 1934 and continuing to be built until 1951, producing an unequalled class of 842 locomotives.

And the 2-8-0 goods engine that hauled the long coal and military trains vital to the war effort, many of which served abroad. Some of which have been repatriated to work on heritage railways today, one on the Severn Valley Railway.

But the emphasis of the thirties was speed and glamour. This, Gresley brought to the LNER with his streamlined A4 Pacifics, hauling specially-constructed coaches in Art Deco style on the London to Newcastle "Silver Jubilee" Express. On a trial run in 1935, his A4 number 2509 Silver Link set a new record of 112mph. Not to be outdone, Sir William (as he became) Stanier at the LMS snatched the record back with 6220 Coronation, the first of a class of that name.

Before launching their Coronation Scot service from Euston to Glasgow, the LMS took a special train of invited guests down the West Coast Main Line on 29 June 1937. Just south of Crewe, where the locomotive was built, Coronation was accelerated to 114mph, just pipping the fastest time set by Silver Link. The driver's name has not entered the history books.

The record has since been disputed, but what is indisputable is the near-disaster that this reckless driving occasioned.

Coronation Scot was going too fast for the crossover points at Crewe and was only saved from going off the rails by late, hard braking that smashed all the crockery in the dining car and no doubt scared the travelling VIPs witless. But the engine was undamaged and "did the ton" for several miles on the return journey to the capital, a testament to the sturdiness of the design.

And the best was yet to come. The zenith of this "golden age" was a new world speed record for steam, one that has never been surpassed. Careful preparations preceded the bid for railway immortality on 3 July 1938. Gresley Pacific 4468, Mallard, was chosen to make the bid for the fastest time on rails. Painted garter blue, and streamlined like her sisters in the A4 class, she became famous as "the blue streak." When one of these unsurpassable locomotives came roaring into Doncaster on the through line in the 1950s, we trainspotters always chorused "Streak! Streak!"

On that fateful day, there were no spectators. With veteran driver Joe Duddington, aged 62, at the regulator and fireman Tommy Bray on the shovel, both based at Doncaster's Carr MPD, the LNER were out to beat not only the existing UK record of 114mph held by an LMS Coronation class, but also the world's best, held by DRG Class 5 of Germany, which had achieved 124.5mph in 1936.

On the historic run down Stoke Bank between Grantham and Peterborough on the East Coast Main Line, officials in the dynamometer car, a special carriage filled with instruments that clocked the exact time and speed with unerring accuracy, registered 120mph, easily seeing off the LMS record. But there was just time before Mallard hit the Essendine curve to go for the big one.

Driver Duddington let her go, and she raced up to 126mph, an astonishing feat that caused mayhem in the short rake of coaches. It is said that the train rocked so violently that dining car crockery smashed noisily and red-hot cinders flying from Mallard's chimney broke windows at Little Bytham. Driver Duddington was forced to brake so hard that the locomotive's big end bearings overheated and she had to be coaxed slowly into Peterborough to avoid deadly damage.

A sign by the lineside now marks the approximate spot where this feat was achieved, showing a metal sculpture of the locomotive with the framed legend Mallard 126mph, 3 July 1938. I often look out for it on the way up to London, but today's Azuma trains zoom past so quickly you can hardly take a glance. It's very satisfying when you can.

The story has it that Gresley believed that his "blue streak" could go even faster and planned an attempt to break his own record with a speed of 130mph, but the intervention of war put a stop to his ambitions. Sir Nigel, as he became, is commemorated by a wall plaque and a fine, more than life-size, bronze statue by the booking office at King's Cross station.

The original design by sculptor Hazel Reeves included a duck by his left foot, to represent Mallard, but this was left off after row within the Gresley Society, described as "possibly the most acrimonious argument in the long pedantic history of the railway hobbyist". A petition to reinstate the bird attracted 3,200 signatures, including Michael Portillo and Vanessa Feltz and former Flying Scotsman owner Sir William McAlpine, but was to no avail. This duck did not fly.

As for Driver Duddington, the man on the regulator who dared the impossible, he retired in 1944 and died in obscurity in 1953, aged 76. He had followed Thomas Duddington, his fireman father, on to the footplate and spent his entire working life with the LNER and then BR, driving the crack expresses on the East Coast main line. His end was not commensurate with his brief moment of fame. Joe was buried in an unmarked grave in Hyde Park cemetery in Doncaster.

Decades passed before Friends of Hyde Park got together in 2021 to launch an appeal for a proper headstone to mark the last resting place of this modest local hero. Such was the support for this overdue recognition that the initial target of £2,000 was reached in less than two weeks and over £7,000 was contributed to the appeal before it was closed.

The money enabled the Friends not only to get a headstone with the names of Joe and his wife Alice, but kerbing and a decorative slab bearing the record-breaking run and a bespoke memorial oak bench with a relief of Mallard where the public can sit and ponder the historic deed of this unassuming man.

At a ceremony in October 2023, a crowd braved the rain to hear speeches and a Christian dedication. "It is difficult for us to articulate just how overwhelmed we have been by the response to Joe's appeal and how bursting with pride we are to now see him so remembered," said the Friends. "We were absolutely blown away." Rather like the record that Joe had smashed, Mallard is her own legend, visible for free at the National Railway Museum in York, where she tempts visitors to touch this mighty machine and wonder "How did they do it?"

Behind the glitz and glamour of the express services, all was not well on the railway. Hampered by restraints on what they could charge and out-competed by cheap lorry transport on short haul deliveries, the Big Four struggled to make a profit. They invested in major publicity initiatives – the Southern appointed the industry's first Public Relations Officer – and more flexible fares, but passenger levels continued to fall: from 1.3 billion journeys in 1923 to 1.2 billion in 1937, a drop of 8%. The cause was not hard to find, indeed it was loud and clear outside the big stations, on the street. Before World War One, there were fewer than 100,000 road vehicles. By 1938, there were 1.8 million cars and nearly half a million lorries, many of them bought cheaply by ex-servicemen with government finance.

Having been compelled by politicians to amalgamate, inevitably the companies looked to government for help. A Royal Commission on Road and Rail Transport, set up in 1931, failed to offer a solution. However, a group of experts under Arthur Salter, meeting the following year, produced the 1933 Salter Report, which was adopted by ministers. Some of the onerous restrictions on the railways were lifted by the Ministry of Transport, which also imposed licensing and safety regulations on the burgeoning road haulage industry.

Furthermore, Chancellor Neville Chamberlain sharply increased excise duty so that all vehicles would pay for the annual Road Fund. He left in place, however, the common carrier obligation, a millstone around the neck of the railways that would not be lifted until 1962.

Two other significant events brought the decade to a close. On the Southern Railway, an enlightened management pushed through large-scale electrification of the system, not just on suburban commuter lines to the capital, but on main lines to the South Coast, laying the basis for financial revival. And in the capital itself, London Transport was created, bringing together the Underground and buses in an integrated combine that took the fight to the enemy: the motor car. Its formation in 1933 was fiercely opposed by the Big Four.

Declaring in 1938 that they were on the brink of bankruptcy, the companies launched a "square deal" campaign to demand the right to fix their own fares and freight rates, without government approval.

With a mischievous sense of humour, as rail historian Christian Wolmar notes, the road haulage industry countered with the slogan, "Give the railways a square wheel!" And there the matter stood, beneath gathering war clouds, until Chamberlain, now Prime Minister, declared war on Germany on 3 September 1939, and the railway became part of government once again.

CHAPTER 5

RAIL AT WAR

GROWING UP IN Railway Terrace, there weren't many books. But one I pored over as a child in mute admiration was *The LMS at War*, by George C Nash. The author was, weirdly, to my mind, noted in brackets on the title page as G.C.N of *Punch*, a magazine of which I'd never heard. It was an official hardback publication, imprinted on the cover with the official company stamp.

It was less than 100 pages long, but illustrated with thrilling colour pictures of LMS ships and trains being bombed, tanks being assembled in railway works and women cleaning mighty Pacifics at Camden shed. Black and white photographs of the war effort like the despatch of goods trains from marshalling yards in the dark stimulated the imagination. That slim volume brought to life the reality of war, unforgettably.

So much so that 50 or more years later I went online to find it, the original family copy having long gone the way of railway lost property. And now I know why Sergeant

Routledge of the RAF kept this 1946 publication. It filled in the six-year gap of his railway service, as employment was still called then. It takes me back, too. Just to hold it is to feel a remote contact with those devastating years, when I was just a nipper, a war-baby born in 1943 in the very front room where that book was kept.

Looking at it now, I read that preparations for World War Two had been under way for "some years" before hostilities broke out. Draft proposals were in place, but after the Munich Agreement in 1938, planning for a switch over to war accelerated and, in the following April, instructions were published that the railway would be once again put under government control. This time, it would be different, unlike the advent of the Great War precipitated virtually overnight by the assassination of a little-known Archduke in a faraway Balkan city.

One mistake never remedied in World War One was the decision not to exempt the railway workforce from joining up. They did so, in their tens of thousands, denuding the system of key workers such as drivers and signalmen and taking craftsmen and engineers from the workshops – which were taken over for the war effort. By 1918, it is estimated that 184,000 railwaymen, 30% of the pre-war labour force, had enlisted. Ironically, as we shall see, this short-sighted refusal to make the railway a "reserved" occupation (as it would be in World War Two) opened the way for tens of thousands of women to get jobs on the railway that had hitherto been a male preserve.

Another big difference in 1939 was the Big Four – the

London North Eastern Railway, the London, Midland and Scottish, the Great Western and the Southern – were more ready for war than the hotchpotch of Victorian companies that were operating in 1914. The system's 37,000 miles of track were in good shape, bolstered by £315million of expenditure in the pre-war decade. As one writer observed, "If Britain's politicians or army was not fully equipped for the outbreak of war, its railways were."

The Railway Executive was back in action, effectively nationalising the system for the duration once again. But the world was a very different place. Unlike World War One, when only a few sporadic Zeppelin raids threatened the system, the railway would be under constant threat of attack by the Luftwaffe, especially after the fall of France and her airfields brought England into fighter range.

As in 1914, the railway was tasked with a huge movement of people on the very first day of hostilities. But this time it was the evacuation of more than a million children and mothers from cities deemed at greatest risk, into the countryside. Once again, the railway was up for it, moving 1.3 million people on 3,800 specials, half of them from London.

Each child carried a gas mask – poison gas attacks of the kind used in World War One were feared, but never materialised – some food, a change of clothing, and wore three identity labels.

The LMS at War recorded, "as they entered the railway stations they marched with a good step, but many of their little faces were hard set trying to be brave. Amongst them were Jewish children from Berlin and one child who had arrived from Danzig only a few hours previously. The railway staff on duty in the stations acted as guide, philosopher and friend to many bewildered little people. The children were leaving their homes and, in spite of many outwardly cheery faces, it was a picture pathetically sad. One ten-year-old girl parting from her mother had said, 'Will I ever see you again?'."

When the anticipated bombing did not happen and "phoney war" ensued for several months, some returned – and then had to be taken back again when the Luftwaffe began normal service in 1940.

The railway's next big challenge was bringing home the army it had embarked in Southampton only months previously. Soldiers of the British Expeditionary Force, plus thousands of French troops, were plucked under fire from the beaches of Dunkirk in May and June 1940. Operation Dynamo ranks as the biggest marine rescue in military history, involving several railway-owned ferries. One, the SS Scotia, a former Irish Mail steamer, was sunk by the Luftwaffe. The ship's purser observed with grim humour, "The sinking itself was not very spectacular. I could not swim at all, but I propose to learn to swim now." With 300,000 fighting men safely back in Blighty, the railway had to perform the same operation as before, only in reverse, involving more than 600 trains in ten days to get them away from danger.

The Home Front was now the front line and the railway was a reserved occupation, compelling those who wished to go and fight to get permission from the authorities. It says something for the patriotism of these workers that 60,000 did so and 3,500 paid the ultimate price. Those who stayed in their post faced a nightmare scenario of risk from the air and even more dangerously, from the rigorous enforcement of "the blackout". Running a railway in the dark is difficult: having to do it without light, or the absolute minimum, probably caused more casualties than enemy action.

With the coming of the Blitz in London and extended bombing of cities like Coventry, Sheffield, Hull, Manchester, Liverpool and Bristol, the Battle of the Railways reached its height. Workers toiled ceaselessly to repair bomb damage, often before the all-clear was sounded, getting the trains back on track. The Germans inflicted very heavy local destruction – some hard-hit places got the distressing description of "Coventrated" – but they never brought the system to a standstill. The bomber may have got through, but the railway carried on.

It was, in the words of the LMS history, echoing Churchill, the railways' darkest and their greatest hour. The scene was of almost indescribable heroism. "The courage of railwaymen became proverbial. They worked until they dropped on to the track and in their tracks. They were wounded, scalded, burned and yet carried on. They flirted with death on countless occasions to save property, passengers and the lives of their own comrades. A number were killed outright at their posts. For at the back of everyone's mind was that primary duty of the railways – the service which knows no stay, no stop – was to keep the lines open and the traffic moving."

Men and women often faced catastrophic situations.

"They had hairsbreadth escapes from falling masonry and live electric cables. They entered warehouses aflame from end to end, the girders red hot and hanging down the walls like bell ropes. They handled unexploded bombs.

"Women telephone operators refused to quit switchboards when their office walls lay in ruins about them. Trains were derailed and lifted bodily from the track, and drivers, firemen and guards crashed with them. Signal boxes received direct hits, huge cranes toppled over, hotel roofs fell in, horses stampeded, goods yards were gutted."

Fire and rescue teams worked night and day in smoke, steam and fire, and it went on night after night as raid succeeded raid, testing railwaymen's ingenuity and resourcefulness – coupled with good honest British grit. That was what my RAF Sergeant father missed, stationed more than 1,000 miles away in the Middle East but he would have agreed with author George C Nash's verdict, "Havoc without plan seemed to be Hitler's principal aim. Terrorise the population and the war was over – an underestimation of the British people and the British railwayman that possibly lost him the war and certainly lost him the Battle of the Railways."

One example of heroism is often quoted: the Soham tragedy. On 2 June 1944, Driver Ben Gimbert and fireman Jim Nightall were hauling a train of 50-pound bombs when fire broke in one of the waggons. Rather than risk the whole train going up and demolishing the Cambridgeshire town, they detached the rear part and carried on with the burning waggon in the hope of reaching safe open country. They didn't get very far. The munitions detonated, smashing every window in Soham. Driver Gimbert survived the blast, but fireman Nightall and signalman Frank Bridges died. The George Cross, instituted by the King earlier in the war, was awarded to both footplatemen.

But the railway was also a centre of more mundane wartime life. Doris Shaw was conscripted for war work while still a teenager, employed in the wages department of engineering company English Electric, working seven days a week.

She takes up the story for *My Lancashire*, "I became one of the voluntary helpers in Preston Railway Station Free Buffet. It had originated during the 1914-18 war when thousands of members of the British Allied Forces had been entertained when passing through.

"The Buffet had been set up in 1939 in a large room at the end of Platforms 5 and 6 by courtesy of the LMS. Run by volunteers, it catered by providing refreshment and a place to rest for the millions of men of the fighting forces together with civilians who passed through Preston.

"The Buffet was open day and night and never closed its doors even during air raid warnings, and there were many. My voluntary shift commenced at 21.00 on Friday until 06.00 on Saturday once a month.

"We had to wear a uniform. Mine consisted of an overall and scarf for my head, and a badge: round, red with Preston Free Buffet embroidered on it in gold thread.

"This badge attracted the servicemen, particularly the Americans, who wanted to start a collection of them. They even offered money to buy the badge. I do recall there was no shortage of chewing gum when the American servicemen were in the buffet."

"Quite often the servicemen stayed in the Buffet all evening, either waiting for a connecting train or had missed theirs. It was at these times in the quieter part of the night they talked of where they were going and their families and if they would ever see them again.

"The quantity of food consumed was enormous. In one week, a typical example being 1,500 loaves of bread, with four hundredweight (cwt) of margarine, over 1,000 meat and potato pies, 116 dozen tea cakes and buns, 500 dozen sausage rolls, one cwt of coffee and 156lbs of tea, sweetened with three cwts of sugar, 230 gallons of milk, 200lbs of jam, marmalade and syrup, 20lbs of cheese, along with 30 dozen lettuce, 30 dozen beetroot and 84 dozen tomatoes, when in season.

"Helping there during the war years, some days it was enjoyable, although busy and other days it was sad to hear the servicemen talking about their friends who would never return."

As the war dragged into its fifth year, the bombing raids diminished sharply, but a new menace appeared capable of causing greater chaos, especially in the capital and southeast of England – Hitler's "secret weapon" of flying bombs, the V1 and V2 rockets.

Remarkably, given the daily carnage, passenger numbers surpassed pre-war levels, reaching 1.3 billion journeys in 1944. The railway played a major role in strategic planning for D-Day, running tens of thousands of trains for the invading army gathered in southern England. Under strictest secrecy, top railway officials were told on 26 March – fully six weeks before the Normandy landings – that "the days of waiting are over" and to begin the railway supply offensive.

They had to be brought into the picture so the huge operation could begin. Only a very few knew the details of the invasion, a secret only broken by an official a few hours after troops had landed on French soil.

As the war drew to a close, my LMS historian wrote, the nation's 600,000 men and women who had got the troops, their arms and equipment through to the ports were amongst the first to taste the fruits of victory, as they quickly went back to their peace-time lives. After everything they had endured "those fruits were sweet indeed".

There was bitter fruit, however, in store for the bosses of the Big Four. Labour won a landslide victory in the 1945 general election, on a manifesto committing the government to nationalisation of the railway. The industry had been shadowing that role twice, in two world wars, and in peacetime it was going to be the real thing.

"The future is to those who shape it, and the time for shaping is now," insisted our LMS historian.

He was assuming this would be a return to the pre-war status quo but this was the eve of Labour taking the railway fully into public ownership.

CHAPTER 6

WOMEN AND THE NEW ORDER

IN THE BEGINNING, this was Adam's world, where Eve had no place. The railway was a man's job, and women were not welcome. It was not that the work was too strenuous for what they called "the gentle sex". The footplate on early small-boilered locomotives scarcely required Herculean strength. But these were the days of Victorian values: the man at his work and the woman in the kitchen (or the bedroom). And so it remained on the railway for almost a century.

For women, the catastrophe of 1914-18 offered a savage irony. With millions of men called to the colours, there was work to be done and not enough men to do it. There had been a very few exceptions to the "men only" rule before, in light work such as operating gate level crossings. Now they would be needed to take on key functions and public-facing roles, working alongside men – if not quite as equals then as fellow railway staff.

There was no direction of female labour as was to come in World War Two, but with almost one in three men – 184,000 – of the railway workforce enlisting by 1918, the industry had no option but to turn to women. They, too, volunteered in large numbers, with 47,000 registering for war duties by the Spring of 1915. Neither managers nor unions were keen to embrace change – this was Edwardian England after all and women did not have the vote – but the Executive Committee bowed to the inevitable and lifted the ban on females some six months after the outbreak of hostilities.

Initially, they were restricted to roles like carriage cleaning and clerical duties, but after conscription for men was introduced in January 1916, they became porters and ticket collectors. As Christian Wolmar observes, "The brave women pioneers who took on these jobs must have found it tough as they had to expose themselves daily to curiosity from the passengers, and even the press." Indeed, the media even suggested women should not be made ticket collectors because they might not be able to handle obstreperous passengers, as if they had no experience of such behaviour in the home.

Their courage prevailed and by the end of the war, they had moved into a variety of male-dominated roles such as engine cleaning – the first rung on the ladder to the footplate. But that was as far as they got. The drivers' union Aslef even opposed that step. In fact, it was not until 1979 that women were allowed to take the ultimate prize. Even in the 1970s, one rail union leader would growl, "If God had meant women to drive trains he would have put sinks in the cab."

Despite it being wartime, women faced discrimination. Company managers were more anxious to have their work than to pay them as much as men. They were denied overtime, annual pay rises given to men and, for several years, the war bonus of three shillings (30p) a week, then only receiving part of it eventually. When peace returned, the men came back to claim their jobs, but women had made the big breakthrough, and the railway would never again be the all-male bastion. The war that made so many of them widows and fatherless had changed their working lives forever.

By 1938, 26,000 women were employed on the railway, chiefly as clerical and domestic staff: clerks, typists, telephone/telegraph operators, cooks, mess-room attendants, refreshment room staff, carriage cleaners and crossing keepers - still less than ten per cent, but a vast improvement. And there was no delay in employing women in the second war. That debate had been won. Their roles were wider, including portering and signalling, but not managerial or supervisory roles and they still got no nearer the footplate than up on a ladder cleaning the engine.

An evocative wartime painting shows three women in blue overalls perched high above the ground washing a stream-lined Coronation Pacific outside Camden shed. A fourth stands by with a bucket, while a couple of men in overalls lounge at the shed entrance, with their backs to the women. The picture speaks volumes about prejudice in the workplace.

That's the same prejudice also recorded in the book Female Railway Workers in World War Two, in which writer Susan Major features the firsthand accounts of women who worked on the wartime railways. Gladys Garlick recalled her training as an LNER guard at Hatfield, "There was a bit of a bad feeling by some of the guards cos they thought it made their job look cheap. Well, I suppose in a way it's like my husband working to be a driver. You don't jump straight into it, do you? You have to work your way up to get to be a guard. And they were, some of them were a bit resentful of that. But on the whole they were all very good."

But there was progress, even in the language of the industry. As the Ministry of Labour began directing women into war work on the railway, the term "railwaywoman" came into usage. By 1943, their numbers peaked at 105,703. And the rules of the game had to change. Traditionally, new entrants came from railway families, with jobs like fireman and driver being handed down from generation to generation. Boys and men were employed in starting grades, working alongside an experienced hand, before moving to responsible jobs.

In wartime, the LMS historian writes, it soon became clear that not enough men were coming through and the railway had to turn to the only substantial source of labour that was available: the "country's womenfolk" and in due course they were engaged in beginners' jobs "in a really big way" and steps were taken to give them every encouragement and assistance to learn the ropes. Eventually, this process had to be accelerated, moving into guards and signals roles. By the end of 1944, the company employed 377 women guards and 623 signalwomen.

They finally worked in nearly 250 railway grades, diversely including concrete workers, sailmakers, assistant architects, fitters, electricians, boiler cleaners, weighbridge men, painters, lock keepers, stablemen and blacksmiths. At the peak, 39,000 were on the books, some 17% of the total.

Among Susan Major's other wartime voices there was Nellie Nelson, an LNER York porter in 1940, who told how she helped to get injured passengers off a bombed train in 1942. And Joan Richards, a GWR parcels clerk at Hartlebury in Worcestershire, who remembered how her father warned her of bad language from male colleagues.

With what now feels like patronising sentiment, the LMS historian wrote – in 1946, remember, "It was all a very remarkable achievement and perhaps one of the happiest features of the new set-up was the readiness with which the regular staff adopted the newcomers, cheerfully helping them over their initial hurdles and displaying a fatherly care in their development." Traditional attitudes returned, however, after the war ended in 1945. Once again, men came back to claim their old jobs and while a higher proportion than 1918 remained, most of the women had to leave. When the industry was nationalised in 1948, only 9.5% of the workforce was female, and those were mostly found in clerical grades or on "light duties". My father Harry, not exactly a feminist, worked with women clerks at Normanton station. One of whom recalled to the family, "I found him courteous and easy to work with."

Nationalisation shifted the focus away from "who does what?" to "who runs the show?" Change had been a long time coming – public ownership had been mooted as early as 1844, in the draft version of Gladstone's Railway Regulation Act, but never acted upon in the face of fierce opposition from the train companies, whose hostility to the idea never truly went away. It resurfaced briefly after World War One, when the companies were financially on their knees, but swiftly receded and didn't re-emerge as a serious proposition until World War Two.

Labour was elected in July 1945 and was true to its manifesto pledge to nationalise the railway, along with coal, steel and the wider transport industry. The legislation took an inordinate length of time, because rail company bosses opposed the policy in principle and fought hard for adequate – indeed, excessive – compensation for their shareholders that saddled the industry for years with mind-boggling interest charges. It was not until 1 January 1948 that British Railways, under the overall suzerainty of the British Transport Commission, a behemoth with responsibility for everything from docks and road haulage to rail and canal, came into being.

It was a moment that the railway unions had demanded for decades, but the newborn was a sickly child. The infrastructure was run down, track and locomotive maintenance had gone to pot and despite rapid demobilisation there weren't enough men to do the jobs. It wasn't even clear how many employees there were: 650,000 was a good guess but it could have been almost 20,000 fewer. Passenger numbers rose sharply during the war, but fell away in the years that followed as fewer could afford to travel and relaxation of petrol rationing made the car more attractive. It was a lose-lose scenario.

The actual shape of the new railway looked remarkably similar on the map. BR was divided into six regions, based on the Big Four private companies that had been dragged kicking and screaming (via the bank) into public ownership. The Western was the GWR virtually intact, and the Southern ditto. The Midland was the old LMS, while the LNER was split between the Eastern and North-Eastern, with a border at Doncaster. Scotland was divvied up as the only "nation" railway, with the LMS having the lion's share. The lion, incidentally, was the new BR logo on its engines, 20,000 of which it inherited, some of them museum pieces dating way back into the 19th century.

Company loyalty was slow to dissipate, and regional managers sometimes behaved like mediaeval barons, defending and promoting their territory. The railway is like that, demanding a fierce allegiance out of all proportion to monetary rewards, which were modest by the standards of the day. It was always a matter of pride to keep the trains running, a sense of duty necessarily exploited during wartime. It was called upon again in the devastating winter of 1947, when the country was under a blanket of ice and snow for many weeks.

The railway gradually settled into a proper national network. Bomb damage was repaired, titled trains reappeared and a fresh generation of locomotives was ordered by Robert Riddles, the new Chief Mechanical and Electrical Engineer ("electrical" had been added since the steam era, a sign of things to come) from BR workshops. They came into operation in 1951 and building continued until 1960, long after the decision was taken to phase out steam as quickly as possible. In all, 999 "standard" locomotives of 12 classes, including the Britannia Pacifics, the hugely-successful 2-10-0 freight engines that were equally at home on passenger trains and a retread of Stanier's go-anywhere Black Fives. Many of these fine locomotives had a very short working life – some only a few years – but no fewer than 46 survive on heritage railways.

BR became a part of the national story, in ways unintended and unwished-for. The British Rail sandwich was portrayed as a dry, curled-up offering with unidentifiable filling, though in fact the industry pioneered cellophane wrapping to keep the product fresh. And the joke originated a century earlier, when novelist Charles Dickens wrote in 1866 of the "British Refreshment sangwich" with its sawdust filling. BR also became something of a byword for lateness, drabness and unhelpful porters.

In its early years, with more money in people's pockets and better holidays, the state-run railway began showing a slight profit. But it was a false dawn. As motoring increased in popularity, and a Tory government denationalised road haulage in 1953, the losses mounted year on year. It became necessary to "do something". The outcome was the ill-fated 1955 Modernisation Plan, a billion-pound scheme to replace steam with diesel by 1970, rebuild the stations and electrify

1,500 miles of main line track. It foolishly promised to win back traffic from the roads and expand freight. Neither happened, and BR was left with massive, white-elephant marshalling yards for goods trains that never ran. Electrification was abandoned, and diesel locomotives proved less reliable than bargained for. Passenger numbers fell, and the closure of unprofitable branch lines began.

Accidents still dogged the railway, no matter who owned it. The ink was barely dry on the legislation in 1948 before 24 were killed at Winsford in Cheshire, blamed on a signalman's error. The 17:40 Glasgow to London Euston train hauled by LMS Princess Royal Class 4-6-2 No 6207 Princess Arthur of Connaught had stopped on the line because a soldier on leave pulled the communication cord – it's thought so he could disembark near his home. A postal express then ploughed into the passenger train. The signalman had thought the line was clear. The soldier who pulled the cord appeared at the subsequent enquiry to confess and explain he had worked as a signalbox lad in Winsford Junction, before the war. It was a job he returned to until he retired in the 1990s.

Almost 40 more accidents occurred during the life of British Rail, taking (by my calculation from official records) a total of 755 lives. Three horrendous pile-ups largely made up this grim tally. The ghastly Harrow and Wealdstone crash of November 1952, killed 112 and injured 340, during fog, when the driver of an express failed to obey signals and crashed into a stationary train and a third ran into the wreckage.

In the Lewisham crash of December 1957, 90 died when signals were missed, again in fog, and a steel bridge fell on the wreckage of derailed trains; and ten years later, 49 died in November 1967 in Hither Green, near Lewisham, when a Hastings to Charing Cross express was derailed at high speed by a broken rail. The Clapham collision of December 1988, caused by a signal rewiring error, killed 35 and injured 484. Thereafter, fatalities were to number in single figures yearly, and the penultimate year before privatisation, 1994, zero.

Safety is always the railway's main concern, but commercial thoughts are never very far behind. As the years of public ownership progressed, competition from the roads hotted up. The first British motorway, the eight-mile Preston by-pass, opened in 1957 and two years later the first 72-mile section of the M1 was opened by the new Transport Minister, Ernest Marples, a road zealot. His family firm Marples Ridgway was in the highway construction business.

After a decade of public ownership, it could be argued that the railway was in a worse state than under the Big Four. Losses soared, the M-Plan ran over budget to £1.5 billion and MPs on the Commons Nationalised Industries Committee demanded that something even bigger had to be done, and it wasn't what they thought would be throwing good money after bad. With road fanatic Marples at the helm, you can guess what came next?

CHAPTER 7

SMASH

BY ITS VERY nature, the railway was always going to produce disaster stories. The nation was adapting from centuries of travel by horse and carriage to a revolutionary system powered by newly-invented steam engines, pulling waggons weighing hundreds of tons, carrying dozens of passengers, travelling at much higher speeds and crewed by a handful of men with little or no safety training and zero regulation. Add in the human factor and you have a perfect recipe for accidents. The only surprise is that there were so few.

Appropriately enough, the first fatalities came on the Stockton and Darlington. An unnamed woman described as "a blind American beggar" was reported killed by a train on 5 March 1827. The first "railway servant" (as the workers became known) to die was fireman John Gillespie, who died when his Locomotion No 5 exploded at Simpasture Junction. The other fireman, Edward Corner, survived the boiler blast and went on to become a motive power manager.

Tough breed, these footplatemen. But that did not help driver John Cree, killed in a copycat explosion of Locomotion No 1 at Aycliffe Lane station only four months later. They were not the last to suffer the lethal unpredictability of early engine design.

More famously, the railway claimed its first high-profile casualty at the opening of the Liverpool and Manchester Railway, on 15 September 1830 when William Huskisson MP was run over by Stephenson's Rocket. Huskisson, former President of the Board of Trade, ignored advice not to climb out of the Duke of Wellington's carriage during his train's coal and water stop. He was hit by a swinging door trying to clamber back to safety and thrown under the locomotive's wheel. He suffered horrific injuries and died a few hours later. Ironically, the national publicity of his fate served to increase public interest in this new-fangled business of long-distance rail travel, the Inter-City of its day.

After this tragic death, the toll of death and injury multiplied as fast as the system expanded and men taken from the fields learned how to manage their new charges. In 1831, PC Bates of the railway police, was fined £3 (equal to nearly £400 today) for causing an accident when he fell asleep and failed to change a set of points. In May 1833, a collision with a farmer's cart in Bagworth, Leicestershire, led to the invention of the locomotive whistle. Three years later, a derailment at Wetheral on the Carlisle to Newcastle claimed three lives. In 1840, five died in another derailment at Howden, east Yorkshire, one more in October in Farington, Berks when the driver fell asleep at the controls and two men died in a boiler explosion in Bromsgrove, Worcestershire.

And so it went on, throughout the decade. A passenger was decapitated in Barnsley. Two died in a head-on collision in fog in Beeston, Nottinghamshire. A bridge collapsed over the Dee at Chester, killing five. Seven more were lost at Wolverton because of a signalling error. An express crash into a cattle van at Shrivenham, Berks, took another six and a derailment in Cumbria caused by axle failure claimed five lives when the train fell down an embankment.

These incidents claimed lives but not much publicity. It fell to the unrivalled pen of novelist and campaigner Charles Dickens to describe the horror of death on the line, after he was involved in the Smash at Staplehurst on 9 June 1865. His train ran over a bridge where rails had been removed for work, and the resulting crash killed ten and injured 49 – the worst to date. In a letter to his close friend Thomas Mitton, the great Victorian wrote, "I was in the only carriage that did not go into the stream... and hung suspended and balanced in an impossible manner." He shared his compartment with two women, whom he composed before climbing out of the window.

"Looking down, I saw the bridge gone, and nothing below me but the line of rail. Some people in the other compartments were madly trying to plunge out of the window, and had no idea that there was an open swampy field 15 feet below them and nothing else!"

He accosted the two guards running down the track, who recognised Dickens and gave him a key to open the wrecked carriage and save the passengers, which he accomplished with the aid of a wooden plank, noting that apart from two baggage vans the rest of the train was in the water.

"Suddenly," he wrote "I came upon a staggering man covered with blood (I think he must have been flung clean out of his carriage), with such a frightful cut across the skull that I couldn't bear to look at him. I poured some water over his face [gathered from the stream in his hat] and gave him some drink, and laid him down on the grass, and he said 'I am gone' and died afterward.

"Then I stumbled over a lady lying on her back against a little pollard-tree, with the blood streaming down her face (which was lead colour) in a number of distinct little streams from the head. I asked her if she could swallow a little brandy and she just nodded, and I gave her some and left her for somebody else. The next time I passed she was dead." A man implored him to find his wife, but she was among the fatalities.

"No imagination can conceive the ruin of the carriages, or the extraordinary weights under which the people were lying, or the complications into which they were twisted up among iron and wood, and mud and water."

Dickens told his solicitor friend that he didn't want to appear at the inquest that followed, or write about it – though his graphic account by letter is the first, and most compelling of its (and perhaps any) time. At his home in Gad's Hill, Kent, a "very quiet" Dickens mildly congratulated himself on his "constitutional (I suppose) presence of mind" at not being in the least fluttered at the time of the crash, but it came to him now. "I instantly remembered that I had the MS of a number with me and clambered back into the carriage for it. But in writing these scanty words of recollection I feel the shake and am obliged to stop."

In the aftermath, Parliament passed a Regulation of the Railways Act in 1871, providing for greater inspection of the companies and their operations, and requiring notification of accidents that caused, or may have caused, loss of life or injury. The new law also provided for accident inquiries to be held, and demanded that companies produce annual performance statistics.

Fatalities fell for a time, but surged again. In 1872, 12 dead at Kirtlebridge, in 1873, 13 at Wigan, in 1874, 16 at Bo'ness, 25 at Thorpe and 34 at Shipton-on-Cherwell. It was a good year when the death rate was in single figures.

As the system grew exponentially, so did the casualties. Drivers blamed the greed and managerial incompetence of the railway companies, who objected to any regulation that might impede the pursuit of profit. The record of accidents due to insufficient planning and gross overworking was, they argued, "simply hideous."

The evidence of one man to the Royal Commission on Railway Accidents 1877 was alarming. Stationmaster Hargraves on the Liverpool and Manchester said, "Drivers sometimes make ten hours overtime at a stretch, and four to five hours is quite a regular thing." Overtime was usually unpaid. In further testimony, a driver named Weston on the North British said he complained to his foreman about the hours. "I have been on duty 16 hours a day in succession and on the third day I went on the engine at 7.30 in the morning and left it at 11 or 12 at night. I took my meals and everything on the engine. I never left the engine. I complained to my foreman and told him that I found difficulty in keeping my eyes open. Upon the third day, I told him that I could not hold myself responsible if anything occurred to the engine or the passengers and that it was unfair to force us to do it. He reported this to the Superintendent, Mr Wheatley, who called me up and said 'Weston, unless you retract those words, I will dismiss you.' I said, 'Mr Wheatley, you have the power to dismiss me, but I cannot retract what I said'." History does not record what happened to driver Weston, but instant dismissal was commonplace. Worse, much worse, was yet to come, on the North British itself.

The collapse of the Tay Bridge in Scotland set a new record of 75 estimated dead. The true number was never established. During a violent storm, in the dark at 19.16 on Sunday, 28 December 1879, the bridge collapsed as an NBR service from Edinburgh to Aberdeen was crossing, plunging the train and all aboard into the freezing waters. At first, incredulity gripped operating staff. The signalman on the south side refused to believe that his train had simply disappeared.

But it had, along with the high girders that formed the centre section. The bridge had only been opened the year before, earning the manager of the North British, Thomas Bouch a knighthood from a Queen Victoria grateful for shortening her journey to Balmoral in August 1879. Some might argue that the over-ambitious, two-mile long link was jinxed. Twenty of the 600-strong workforce had died during construction. Certainly an official inquiry found many faults in the design, particularly weakness of the cross bearings of the piers, and their inability to withstand gales prevalent on the Firth of Tay. Bouch himself was found largely responsible and died in disgrace less than a year later.

Astonishingly, given that not all the victims' bodies were recovered, the locomotive was. NBR 224, a 4-4-0 only eight years old, was salvaged and repaired and continued in service for another 40 years. Drivers gave her the macabre nickname of "The Diver" and some were reluctant to take her over the new bridge, completed in 1887 and built alongside the stumps of the old one, which are still visible today.

Had the awfulness of that day's events not been sufficient to ensure their immortality, William McGonagall's epic poem *The Tay Bridge Disaster* would certainly have been. His "so bad it's good" poetry has inspired generations of comedians, imprinting the catastrophe on the popular imagination and not just in Scotland where it is ruefully revered.

The carnage continued with 25 dead at Hexthorpe, Doncaster, in 1887, when an express ran into a stationary race train. Perhaps the least accidental accident took place at Thirsk, north Yorkshire, in November 1892, when ten died in a rear-end collision caused by human error: the signalman deprived of sleep by his ill and dying daughter was ordered to work even though he declared himself unfit and explained why. He made a mistake, but was convicted of manslaughter. The court gave him an absolute discharge.

The accident rate – or, at least, the toll of dead and injured – subsided in the new century, averaging about one a year. Still, 16 died when a train ran into the buffers at Glasgow St Enoch station in 1903, seven more in a fog collision at Cudworth, West Yorkshire in 1905, 20 at Blundellsands,

Lancashire when two electric trains collided. A collision at Salisbury in 1906 took 28 lives and 22 more perished in a blizzard in Forfar and 18 in a derailment at Shrewsbury in 1907. And those were just the worst of that decade.

But despite the intense overcrowding of the system during World War One, there were relatively few serious fatalities: except the aforementioned Quintinshill disaster. Peace in 1918 brought reduced activity on the railway, and something of a let-up in the bloodshed, with only seven dead until 1921, when a head-on collision on a single line at Abermule, Montgomeryshire, blamed on inattention to proper procedures killed 17 and injured twice as many more. Three died at Birmingham New Street in November 1921, in a collision caused by driver error and "wet lines". And that was about it in the expansionist years of the British Railway. The companies that had created the greatness, and killed so many in the process, were about to die themselves.

But the carnage of accidents continued after they grouped into the Big Four. Three died at Retford in February 1923, blamed on driver error, four at Diggle that November, from the same cause. In November 1924, 15 were killed in a crash at Lytham, due to a fractured rail tyre and nine the following year at Fenny Stratford following a collision with a road vehicle and nine more in a similar accident in November at Parkgate, Rotherham. In 1927, a head-on collision at Hull Paragon station took 12 lives, a crash at Sevenoaks, 13 and another head-on took 25 at Darlington in June 1928, and 16 more at Charfield, Gloucs, when a gas-lit night mail passed 'signals at danger' and burst into flames.

Despite higher speed, the thirties began with fewer disasters. But 12 died near Warrington in September 1934 due to a signalling error, 13 the following year at Welwyn Garden City, with the same cause. The hideous Castle Cary, Somerset, crash in "whiteout" in December killed 35 and injured 179.

World War Two, with the railway operating under blackout and the threat of Nazi bombing, could have been catastrophic, but accidents were no more frequent than in peacetime. Some of those that did take place, however, were pitiless. In November 1940, 27 died when a driver took his train off the rails at Norton Fitzwarren; nine at Holmes Chapel in September 1941 (signalman's error); 23 at Eccles the next month, blamed on fog and the blackout; 14 at Cowlairs, Glasgow, in February 1942, signalman's error; nine, including World War One hero and Conservative MP Colonel Frank Heilgers, at Ilford in January 1944 in fog.

Considering the operating conditions of wartime, it could have been much worse. Peacetime brought the Bourne End disaster of September 1945, with 43 fatalities, blamed on excess speed, 20 more died at Lichfield in January the following year after signals froze and a further ten the same month at Browney near Durham when a main express hit a wrecked goods train. 1947, the last year of private ownership, was appalling: nine dead at Gidea Park in January (signal error), six in July at Polesworth (track defect); 18 in August at Doncaster (signalman's error); 32 dead and 183 injured in October in the infamous South Croydon smash in fog due to signalman's error; 27 more the same month at Goswick,

Northumberland, due to excessive speed; four dead and 101 injured at Motspur Park in November (fogman's error) and finally one more in December, due to a runaway fuel train at Manchester Victoria. What a way to end the era of private ownership.

CHAPTER 8

DR BEECHING'S SURGERY

IF THERE IS one name that strikes terror into the hearts of railway lovers it is that of Dr Richard Beeching. This smiling, toothbrush-moustachioed, avuncular-looking man was responsible for the reshaping of the system, some would say virtually out of existence, more than 60 years ago with his infamous Report.

All the other reviews, investigations and policy papers since nationalisation have been forgotten. Who now remembers the Serpell Report, or the Guillebaud Report? Buried, bygone. But not the portly physicist of the sixties. His name is even perpetuated today in the Signals Failure column of satirical magazine *Private Eye*, under the pseudonym Dr B-Ching. That's fame for you!

Nobody could have known that such lasting notoriety was in store for the boy born in April 1913 in Sheerness, on the Isle of Sheppey, the second of four brothers to Hubert Beeching, a reporter on the *Kent Messenger*, and his schoolteacher wife. He attended All Saints C of E primary school in Maidstone and won a scholarship to the local grammar school – a path I was to tread 40 years later.

A brilliant young scholar, he went on to take a First in Physics at Imperial College, London, before doing a PhD and then working on armament design during World War Two. He was obviously too valuable to be wasted in the army (his brother Kenneth was killed) and after the war he joined ICI as a high-powered technical assistant, rising quickly to the board of the Fibres Division. In 1956, he went to Canada to run a terylene plant in Ontario, before returning as ICI technical director.

Nothing, so far, marked him out as the butcher of the branch line, as he became known to railway people. But in 1959, he accepted an invitation to join the Stedeford Committee, set up by Tory Transport Minister Ernest Marples, to advise on the financial state of the British Transport Commission, the umbrella body overseeing all state-owned transport including the railways. It might not have been explicit in the committee's terms of reference, but the idea was to dismantle the "socialist" BTC and make the railways pay their way. In March 1961, Marples appointed Beeching the business-men who knew about paint and terylene – and the bottom financial line – but rather less about trains and the people who run them, as the first chairman of the British Railways Board. He was to keep his ICI salary of £24,000 a year (more than £500,000 today), a decision that astonished the media and possibly angered other nationalised industry bosses who earned less than half that.

It was not an auspicious start to a controversial career as Mr Railway. Beeching took over a system in dire financial straits, with losses almost tripling to £42million in 1960. The much-vaunted Modernisation Plan of 1955 was a costly failure, with the creation of white-elephant marshalling yards, a botched transformation from steam to diesel, on-off plans for electrification and unhappy industrial relations. Frankly, it was a mess, and Beeching's business brain took one look and decided that radical surgery was critical. That, indeed, was his own epitaph, "I suppose I'll always be looked upon as the axe-man, but it was surgery, not mad chopping." That depends on how you define surgery. Beeching took off too many limbs.

As instructed by road zealot Marples (he owned a construction company that built motorways), he published his first report in 1963, modestly titled 'The Reshaping of the Railways'. Identifying "the good, bad and the indifferent" (where was "ugly"?) it showed that losses were on an unimaginable scale. The least-used 50% of stations contributed only 2% of passenger revenue, and one-third of route miles carried just 1% of passengers. Beeching proposed closure of a third of the country's 7,000 stations and 5,000 miles of track, including some main lines, with the aim of saving £18million a year to bring the railway into profit by 1970. To the consternation of the unions, 70,000 jobs would be lost.

This, in an industry that had traditionally offered "a job for life", albeit not well-paid but secure, often passed down from generation to generation. The railway family was never the same again, like the railway itself.

The Beeching Plan was just what his political masters had wanted, and the Tories began implementing it with a vigour. An "orgy of closures" followed, 2,363 stations in total, taking no account of the social consequences for people living in the country served by "uneconomic" branch lines, or the extra traffic heaped on the road and increased traffic accidents. His alternative strategy for "bustitution", replacing trains with buses, proved a mirage.

Harold Wilson's Labour Opposition fiercely opposed the cuts, a popular policy that helped him win the October 1964 general election, albeit with a tiny majority. Not daunted, Beeching returned to the operating theatre in February 1965 with a second, even more radical, report. His new vision was for traffic confined to nine key routes of 3,000 miles that would be "selected for future development". One casualty of this downgrading would be the east coast main line north of Newcastle. Shades of the Race to the North on scenic cliff-top tracks!

This was too much for the incoming Labour government. His plan was rejected and the good doctor was invited to return to his old job at ICI. But his original remedy for the railway was accepted. In a spectacular U-turn, Wilson and his lacklustre Transport Minister Tom Fraser, announced that they had no power to reverse the Beeching closures and proceeded to intensify the programme, shutting down 1,071 miles in 1965, more than the hated Marples had the year before.

Not every line marked for the Beeching scalpel was knifed. The Central Wales Line from Shrewsbury to

Swansea, passing through thinly-populated mountains escaped, largely because of pressure from MPs through whose marginal constituencies it passed. The Far North and Kyle of Lochalsh lines survived in part for military reasons. The Hope Valley route from Sheffield to Manchester, and the line to Ilkley, were reprieved and flourish today. The Oxford to Cambridge "Varsity line" serving the fast-growing new town – now city – of Milton Keynes succumbed and is now planned for reinstatement, at a cost of billions.

The savings were never on the scale envisaged, only £30million a year, when losses were mounting to £100million. It was left to a later, more politically-committed Transport Minister, Barbara Castle, to pick up the pieces with legislation that recognised the social value of railway services, with appropriate subsidies from the public purse.

Beeching, still only in his 50s, didn't look back in anger. Wilson rewarded him with a life peerage in 1965, creating Baron Beeching of East Grinstead, where he lived with his wife Ella, his childhood sweetheart who he married in 1938. A man rarely photographed without a smile on his face, even while smoking expensive cigars behind his desk, he appeared unperturbed by his political defenestration. He rose to become deputy chairman of ICI, and a director of Lloyds Bank. But if he had finished with public life, it had not finished with him.

Harold Wilson brought him back in 1966 to chair a Royal Commission on the courts, which resulted in a radical restructuring of the system with regional courts in the big cities under an Act of 1971. This time, no one accused him of butchery.

Beeching died quietly in Queen Victoria Hospital, East Grinstead, aged 71. There were fewer obituaries for him than for the branch lines he closed, but he left behind a legacy of controversy that persists today. Some writers seek to exonerate him from guilt, arguing that the politicians of both major parties were to blame for the partial destruction of the system. Enthusiasts maintain that the "Beeching Axe" was madness. At this distance from the war of words, it is possible to say that the truth probably lies somewhere in the middle. The process that he accelerated had been going on for decades.

Even before World War Two, some loss-making lines had been closed, and others followed in the fifties after nation-alisation. Railways were under heavy commercial pressure from the motor car and the lorry, unable to get rid of the "common carrier" tariff that obliged them to take any load to anywhere at a fixed price, rather like the Post Office stamp obligation. Road haulage had been denationalised in 1953, flooding the market with tens of thousands of go-anywhere, charge-anything lorries. By 1960, one in nine households owned or had access to a car and, as has been wittily observed, the other eight were saving up for one. Moreover, road-building, with Marples as its champion in government, was regarded as an investment whereas money for the railway was a subsidy.

The difference with Beeching was that his was a concerted, national programme, designed to change the industry once and for all, making money rather than devouring it. This, as the doyen of railway writers, Christian Wolmar, puts it, meant that he was "searching for that Holy Grail, the profitable railway", something never achieved on the continent or in the USA. On BBC News in 2008, Ian Hislop, editor of *Private Eye* that runs the B-Ching column, suggested that history had been unkind to "Britain's most-hated civil servant" who had been used by politicians to do their dirty work for them. He described Beeching as "a technocrat who wasn't open to argument to romantic notions of rural England or the warp and weft of the train in our national identity. He didn't buy any of that. He went for a straightforward profit and loss approach and some claim we are still reeling from that today."

I hear echoes of my father in the character of the railway surgeon. One of the jobs he had after moving on from the booking office was to go round stations in the West Riding doing efficiency audits. In those days, the platforms had a whitewashed border to warn passengers not to get too close to the edge. This job was done by porters, when they felt like it. My father commented how frequently he found them at this task on his visits, drawing the obvious conclusion that they hadn't enough "proper" work to do.

Beeching must be the only railway figure of modern times to be so often immortalised in public signs. There is a Beeching Drive in Lowestoft (which he would have axed from the network), Beeching Close in Upton, Oxfordshire, and another in Countesthorpe, Leicestershire, and a Beeching industrial estate in Alford, commemorating the station lost under his cuts. In his home town of East Grinstead, locals with a sense of humour wanted to name a former railway cutting converted to a relief road as the Beeching Cut, but some jobsworth insisted it had to be Beeching Way. And the Railway Inn in Aberystwyth was renamed Lord Beeching's for some years. It is now the Hoptimist because they want to bring people together, not isolate them.

Isn't that what Dr B wanted to do, only on a smaller scale, with fewer lines and people? This conundrum is not solved in the episodes of *Oh! Doctor Beeching*, a 1990s BBC sitcom set in a fictitious rural station named Hatley and filmed at Arley on the heritage Severn Valley Railway. The cast was headed by Jeffrey Holland as the stationmaster and Su Pollard as the ticket clerk, who sang an updated version of the 1892 music hall song and referencing the *Oh, Mr Porter!* Will Hay film of 1937: "Oh, Dr Beeching what have you done? There once were lots of trains to catch but soon there will be none I'll have to buy a bike 'cos I can't afford a car, Oh, Dr Beeching, what a naughty man you are!"

Initially popular, audiences for the sitcom declined sharply, rather like real life on a rural branch line, and after two years it was axed, a neat point that the inscrutable Dr himself would no doubt have appreciated.

CHAPTER 9

THE END OF STEAM

THE ACCOUNTANTS, THE public timetable planners and the managers of diagrams – the description of the day, and to this day, of train movements – must have heaved a sigh of relief. The dirty, mechanical, temperamental era of steam was over. But 11 August 1968 was a time of mourning for traditionalists who had grown up with these clanking, smoking monsters and loved them. Something of themselves had gone with the locomotives that shaped their fascination with the railways. And some of the footplatemen shared that sense of loss, deciding that the clean new regime was not the way of life for which they had volunteered.

It had been a long-drawn whistle, like one of these night-time shots on the prairie in American movies, where steam had already shared the fate of the buffalo. But the new breed of traction had been coming for decades, starting with electric locomotives for the North Eastern Railway in 1915, for coal trains on the Shildon-Newport line in County Durham. The experiment lasted just 20 years before it was abandoned and steam returned. Most of the freakish-looking locomotives were scrapped in 1950 after years in storage and the last, named Denis after its driver Denis Dodridge, was withdrawn in 1964 after working on electrification in Essex.

The LMS is thought to have been the first to try diesel traction, manufacturing a prototype shunter in 1931, but it was the Great Western, always in the van for development, that brought out the first passenger vehicle, a streamlined railcar, introduced in 1933, chiefly for work on secondary lines. The class eventually numbered 38 and did sterling service into the 1960s and three are preserved.

Great Western also experimented with gas turbine traction, with two locomotives that went into service in the late 1940s, ordering two prototypes that spent their short lives working expresses from Paddington. One, number 18000, withdrawn in 1960, is preserved at Didcot GWR museum.

The LMS was the first with main line diesel locomotives, building a pair numbered 10000 and 10001 at Derby works in 1947, months before the company was swallowed up by nationalised British Rail. This duo was regularly at work on West Coast main line expresses, always in tandem, in the fifties – I remember seeing them at Lancaster, as a lad of about ten, and being thrilled by their growling power, a sentiment I was eventually to shed. The pair worked on the West Coast main line until the mid-sixties before being withdrawn and scrapped. Enthusiasts are working to recreate 10000, the "ancestor" engine of all we have today.

And by the late fifties, diesel multiple units were operating on a number of lines in Yorkshire's West Riding. Passengers could share the thrill of the cab and the driver's view of the track, by sitting behind his cab, openly visible through a glass window. No hint of a protective screen, though he could let down a blind if we were annoying him. The future – actually, more like the present – was coming down the track. But not without a few stops on the way.

The Western Region was the first to dispense with tradition, when 7029 Clun Castle hauled the last steam train out of Paddington, to Banbury. A rogue survivor, pannier tank 9641 was still active on Croes Newydd shed, Wrexham, in November 1966. The Southern followed suit in July 1967, when 35050 Elder Dempster Line, took a Weymouth train from Waterloo. In Scotland, the last working J36 goods locos dropped their fires in Fife sheds in June the same year. Steam ended in the North-East, where it began, in September 1967, but hung onto the famous Leeds Holbeck MPD for another month. Normanton, my "home shed", continued as a service steam coming over the Pennines. And it was over there that the last act was played, in some style. Riddles's great Britannia Pacifics could still be found fighting the elements on the Settle-Carlisle "roof of England" route in the winter, but work for the few remaining depots tapered off in early 1968.

Lancashire was the last redoubt, with depots in Liverpool, Stockport and Lostock Hall (Preston), Rose Grove (Burnley) and Carnforth serving freight and occasional passenger trains. The buffers were in sight for commercial traffic, but lo, the commemorative railtours began even before sentence had been carried out, starting – where else – with Flying Scotsman in March 1968. The special train, organised by Williams Deacon's Bank Club was so oversubscribed that a relief ran behind the last Standard Pacific still available for work, 70013 Oliver Cromwell.

The last steam-hauled Belfast Boat Express left Manchester Victoria in May, shining black-bright after 100 enthusiasts descended on Carnforth MPD to scrub her up, much to the annoyance of the local police, who detained a few. Some freight workings continued through the summer, but Andy Hall, the last driver on an official BR trip, took 45156 into an empty Lostock Hall shed on 4 August 1968.

They thought it was all over, but in reality it was just a new beginning. A week later, British Rail cashed in with train 1T57, the now-famous Fifteen Guinea Special, billed as the final passenger train to be run over the national network. Every inch of the route was mobbed by lineside photographers and those who just wanted to be part of railway history. Its tour of the North-West was a sell-out, rather like the impetuous end of steam traction by BR, which had only ceased building locomotive engines eight years previously, when 92220, Evening Star was outshopped from Swindon works.

The name for Britain's last locomotive was chosen by three Western Region workers, a driver from Aberystwyth, a clerk in Paddington and a boiler washer in Old Oak Common MPD. And fittingly, it was ceremonially named by an ex-Swindon apprentice, Reggie Hanks, who said proudly

"There has to be a last steam locomotive, and it is a tremendous thing that it should be built here in these great works at Swindon.

"It has been truly said that no other product of man's mind has ever exercised such a compelling hold on the public's imagination as the steam locomotive. No other machine in its day has been a more faithful friend to mankind nor has contributed more to the growth of industry in this, the land of its birth, and indeed throughout the whole world."

Emotional words, hyperbolic even (what about the harnessing of electricity?), but understandable. And modern views of the end of steam criticise the virtual panic with which the process was accelerated. BR chiefs ordered no fewer than 26 different, untested classes of diesel locomotives and, as a result, writes Wolmar, they soon had hundreds of unserviceable engines whose frequent breakdowns attracted much unfavourable press coverage. The passengers were none too pleased, either.

Of course, in the hyper-pedantic world of railway enthusiasm, it is never wise say something is the first, the oldest or the last. So it was with steam. A solitary pannier tank of GWR origin continued to potter about the London Underground, a relic of the days when steam held sway on the Metropolitan line until the late fifties. The final commemorative journey of L94, originally 7752 when it worked in the south Wales coalfield, from Barbican to the sheds at Neasden took place in June 1971. The locomotive was preserved, and returned to the Underground for yet another "goodbye" special in 1980 and is based at Tyseley steam centre in Birmingham – and fit for operations.

What's more, in their haste to nationalise and rationalise, the politicians forgot the last outpost of the old ways, in a remote Welsh Valley. Here, the Vale of Rheidol, was a remnant of GWR days, now part of BR. This narrow – one foot, one and three quarters – gauge line ran for just under 12 miles from Aberystwyth to Devil's Bridge. It was originally built in 1902 to carry pip props for mines in South Wales and lead ore down to the sea, but public demand to ride on the little train soon overtook goods traffic. The line closed during World War One, but survived to reopen and was absorbed into the GWR and finally BR Western Region.

Power came from a small fleet of necessarily diminutive locomotives, built at Swindon, latterly sporting a copper-capped chimney and bearing the BR Rail Blue livery with an Inter-City emblem. With the advent of privatisation, the V of R was sold to investor enthusiasts, and continues in operation today, its four stations and five halts served by three locomotives, Nos 7 Owain Glyndwr, 8 Llywelyn and 9 Prince of Wales. The line has appeared on TV many times and attracts thousands of tourists every year.

That proves my point: never offer the last word in railway matters, because someone will always say "but you forgot about..." And unforgettably, a total of around 375 steam engines from BR days did survive and are either running today on heritage lines or rusting quietly in a queue for restoration. Indeed, some are running on Network Rail lines, at the head of special excursions that normally attract full bookings. The age of steam is not dead, it has just been taken out of state hands.

The process began early, in fact years before the Fifteen Guinea Special announced the end of an era. The Railway Preservation Society was created in the early 1960s and held its first meeting in 1962 chaired by Captain Bill Smith. He was the first person to buy a steam locomotive from state-owned British Rail. In 1959 he saved the former Great Northern Railway 1247, a J52 class 0-6-0 saddle tank from scrap, kickstarting a revolution in the movement. As renumbered 68846 after nationalisation, the engine had a chequered life. Built by Sharp Stewart in Glasgow in 1899, and first shedded in Doncaster, she ended her active life at Top Shed, Kings Cross, on shunting duties. She was affectionately known as Old Lady. Captain Smith donated the locomotive to the National Railway Museum, and after a Cook's Tour of preserved lines, that's where she now resides.

It was personal initiative of this sort, plus a large dollop of good luck and serendipity, that saved hundreds of steam engines that would have gone to the cutter's torch. In April 1961, the railway press published a list of 71 locomotives earmarked for preservation, many of them already museum pieces. In a story worthy of a *Boys Own* annual, four west London schoolboys, Angus Davis, Graham Perry, Jon Barlow and Mike Peart, decided to buy a GWR 14XX 0-4-2 tank engine to avoid the disappearance of the entire class of this much-loved type, often found on rural branch lines. In a letter to *Railway Magazine*, they successfully appealed for £1,130 in funds and within months the Great Western

Preservation Society had been established. It became firmly established at the decommissioned Didcot motive power depot, which today attracts many thousands of visitors a year and provides servicing facilities for visiting engines.

The preservation picture would also have been very much poorer without the unlikely role played by a South Wales scrap business. The Woodham Brothers, based in Barry near the seaport from which so much Welsh steam coal had been exported, were in the profitable business of scrapping old BR waggons and locomotives, begun in 1959. Hundreds of both stood in forlorn lines, rusting in the salty sea air. Cutting up of the dozens of engines arriving with the rapid end of steam began in 1961, but in autumn 1965 the firm shifted attention to the endless stream of freight waggons and brake vans.

In all, Woodham Brothers bought 297 locomotives for scrap, and the survival of all but 80 in a "steam graveyard" spurred preservationists to intervene with their cheque books. One by one, they were "claimed" by different groups, earmarked for purchase.

Midland Railway 0-6-0 4F freight engine 43924, was the first to be bought in September 1968, destined for the embryo Keighley and Worth Valley Railway. Barry Scrapyard became a cause celebre, even taken up by MPs like puffer nutter Robert Adley, influential chairman of the Commons Transport Committee. One by one, the rusting hulks, sometimes stripped of key parts like coupling rods, and without tenders, were saved from the yard: a grand total of 213 in the years to 1987. Nothing like Operation SaveSteam happened anywhere else in the world.

And not content with this extraordinary achievement, preservations turned their attention to "the ones that got away" – favourite classes that had gone out of existence. So began the "new build" movement, raising millions of pounds to bring back what had been lost. At the time of writing, dozens of new builds are under construction or in service, ranging from the mighty LNER A1 Thompson Pacific, 60163 Tornado, Standard Clan Class 72010, Hengist and LMS Patriot 45551 to a humble 2 Mixed Traffic tank engine 84030 and a recreation of G45 0-4-4 Worsdell tank 1759, dating from 1893. Enthusiasm knows no bounds and the rebuild movement is serious business, employing hundreds in specialist workshops, on the heritage railways and in firms as far apart as Darlington and the Forest of Dean. As Mark Twain said of his prematurely-published obituaries, rumours of the demise of steam have been greatly exaggerated.

CHAPTER 10

RAILWAY COMMUNITIES

THE RAILWAY SHRANK the world, bringing people together more powerfully than any other invention until the radio almost a century later. It did not only connect towns and cities. It took the huge urban population into the countryside and spurred the flow of agricultural labourers from the fields into the factories and the "dark, Satanic" textile mills. It opened up the sleepy seaside to trippers and holidaymakers, particularly after the introduction of paid holidays.

But the railway also created its own communities, company towns dedicated to the industry, to satisfy their insatiable appetite for skilled and loyal labour. They were usually built on greenfield sites where land was cheap, close to an operating centre and often close to the industries they served like coal and iron, or alongside their own engine and carriage manufacturers such as Crewe and Wolverton. Their history is still with us today, in places like Derby, Swindon, Eastleigh and Doncaster.

My own native town of Normanton owes its existence to the railway. It is a settlement of great antiquity, appearing as Normentune in the *Domesday Book*, having a church, a priest and a number of villains (villagers, in this context). In 1822, it was still an agricultural village, with a population of 280. By then, it had acquired a grammar school (which I attended), a public house, the White Swan, a corn mill, a blacksmith, a shoemaker and a wheelwright.

The coming of the railway, the result of a deal between "Railway King" George Hudson, the Mayor of York, and George Stephenson, acknowledged "father of the Railway", turned Normanton into a boomtown. Work began in 1837, under Stephenson's supervision, bringing three companies together to give a direct link from the North to London via the Derby to Leeds connecting south through Birmingham and Rugby, and north to York. It was all change at Normanton! Hudson's dream of a direct link to the capital from his vast business interests in Yorkshire was realised. It did not last long, unlike his notoriety.

The railway birthed engineering legends but it also created unscrupulous, power-crazed business bandits, of whom Hudson, is the most infamous. Born in the East Riding farming village of Howsham, in March 1800 and raised in poverty by his elder brothers, he apprenticed as a teenager to a draper's in York, eventually marrying the owner's daughter and building the biggest business in the city. He grew up a dissenting Methodist but traded up to Tory High Church and moved into railway finance.

After a chance meeting with George Stephenson in

Whitby, he planned the rail connection from London to the North-East, using the York and North Midland Railway of which he became chairman in 1837, the same year he became York's free-spending Tory Mayor. He was already playing fast and loose with the company's money and buying up and then closing competitor lines.

Becoming Tory MP for Sunderland only spurred his insatiable acquisitiveness, accumulating railway companies like a child in a sweet shop. His greed was his downfall. Pilloried as "the Bubble of the age" in a pamphlet published in 1848, his corrupt business practices and dishonest sharedealing were exposed and he was forced to resign from his chairmanship of the York and North Midland in 1849.

Thereafter, Parliament set up a committee of inquiry, which found him guilty of abusing the confidence of investors and wielding power in his own interest. He was charged with bribing fellow MPs and embezzling £600,000 from his own companies. Defeated in the election of 1859, he lost his parliamentary protection from imprisonment and, humiliated, was forced into exile in Paris, allegedly living in shabby hotels and going without food.

Nothing if not shameless, Hudson returned to Whitby five years later, seeking election as MP for the town where he had substantial business interests. On the eve of the poll, he was arrested by the Sheriff of York and thrown into the city jail but released after his debts were paid by a friendly colliery owner. But he was still a wanted man and exiled again until the passing of the 1870 Act abolishing imprisonment for debt and he could return to York after his creditors abandoned hope of getting their money back. He died, a broken man, aged 71, in December 1871.

Critics reviled him, but investors were only to believe the fairy stories of get-rich-quick he offered at the height of the Railway Mania. A fairer judgment may be gained from Christian Woolmar who writes, "Hudson is often denigrated as more of a crook than a railway pioneer, but this is too harsh a verdict on a man who understood the importance of melding the haphazard collection of railways into an integrated role." Once feted by prime ministers and royalty, he now lies in the grave at the church of St Peter and St Paul in the parish of Scrayingham near where he was born.

Back to Normanton, where the station, designed by George Townsend Andrews, was opened as a three-company interchange by the North Midland on 13 June 1841.

At the time, it was the world's longest railway station, boasting a platform a quarter of a mile (400m) in length. For the next decade it was the busiest station in the country, employing more than 700 people and handling 700,000 passengers a year.

Rebuilt in 1871, it had a glass canopy supported by iron pillars, refreshment rooms for the hoi polloi and an elevated, covered walkway over the tracks to the Midland Hotel, where the gentry could take their ease. Express trains for Scotland stopped here for 20 minutes for a refreshment break. However, it soon developed a rather unsavoury reputation. In 1844, the *Railway Times* reported that the porters were rude to all passengers and especially to those coming from Manchester. Unattended single ladies were often put in a state of near-hysteria at the thought of using the station!

This was also an important exchange centre for the Post Office, with mail coming from all points of Britain for direction to its destination. From 1855 there was a full-time GPO site and I remember barrows heaped with bags of mail on the platforms.

Notable visitors in its heyday include Queen Victoria, Prime Ministers Gladstone and Disraeli, former US President General Ulysses Grant and the Emperor of Brazil, Dom Pedro II. The latter appeared on 7 August 1871, at the invitation of Henry Briggs, owner of the Good Hope pit in nearby Whitwood. His imperial majesty was taken on an underground tour of the mine and then to inspect a new shaft being sunk at Loscoe. This new colliery venture was named after Dom Pedro and remained so for more than 100 years. The entrepreneurs then went on to inspect Normanton Iron Works, where the emperor placed a large order for railway lines, before taking the express to Sheffield.

All in a day's work for the small army of porters, servants, signalmen, clerks, foremen and the stationmaster, no doubt, but this was not the true purpose of this proud railway operation, at its height employing 900 people.

Normanton was ringed by at least half a dozen collieries, with many more only a few miles away and capacity for handling the traffic grew immensely to meet demand. Eventually, 13 miles of sidings could accommodate 2,500 waggons. Railway tycoon George Hudson envisaged this site as "the Crewe of the Coalfields" and for the next century it fulfilled that purpose, with miles of marshalling yards for the collection and despatch of coal trains. At Normanton's most prosperous time, the pits and the railway employed 70% of the local workforce.

There had always been coalmining in "Normy", as the locals called it. Records exist of "sea coal" in the adjoining village of Altofts and of mining in about 1600. But it was the coming of the railway that galvanised the coal industry. Numerous pits were sunk after 1840, attracting migrants from all over the country and Ireland. The population soared from 283 in 1837 to more than 10,000 in 1891. So many Irishmen flocked to the St John's colliery, that had the cross of the saint on its coal waggons, that it was known as "the Catholic pit".

It was situated on the fringe of the town, where later a Catholic school and church were built, a testament to the lasting attraction of the railway settlement. Where the mines opened, other industries followed. An ironworks was in production by 1870, and three brickworks opened in the ancient hamlet of Newland next door to St John's using the abundant local sources of clay.

In my class at Normanton Grammar School in the 1950s only one boy, a solicitor's son, could have been described as middle class. "The Forge", a group of terraces near the railway line built on the site of the ironworks that once employed 400 men, was home mostly to miners. The area is now a supermarket. Most people lived in railway housing or council houses, built on estates over the years. One street was named after a train driver in our terrace, Alf Ripley, who was also a local councillor. Labour, of course. The urban district council was dominated by Labour, and the constituency had a Labour MP since the party began fielding parliamentary candidates in 1906 – and Lib-Lab before that going back to 1885.

Normy's prosperity began to diminish with the decline of the railway and the rundown, and eventual closure, of the coal mines. Inter-city trains like the St Pancras to Glasgow, Thames-Clyde Express, ran through the station until the 1950s, but didn't stop. Gradually, the direct services to York, Manchester and Sheffield disappeared. Multiple diesel units introduced in 1960 restored some of the growing commuter traffic into Leeds, but the writing was on the wall.

And then they took the wall down. Normanton station buildings were demolished between 1979 and 1984, and it was turned into an unstaffed halt. Passengers had to walk across the line on sleepers to get to the road, a shortcoming later rectified by a steel footbridge.

Coal traffic held up for some years, but that, too, began to fall away as the seams locally were exhausted. One by one, the pits closed, rendering the marshalling yards obsolete. Snydale colliery closed in 1965, West Riding two years later, Whitwood in 1968 and the last, Sharlston, ended coal output in 1993. The engine sheds – there had once been two, though not in my lifetime – staggered on until January 1968, almost the end of steam. They were a forlorn sight at the end, empty, lifeless, no bustle or men reporting for work, no sounds of locomotives moving off shed to the giant concrete coaling tower.

There were many other "railway towns" across the

country, though few were as dependent as Normanton. Prime examples with a rich place in history include Crewe, Doncaster, Swindon and Derby, whose contribution has been amply documented. But they sometimes sprang up in the most unlikely places, and that is where my interest takes me.

Take Newton Abbot in Devon, the most westerly. First documented in the 12th century as Nova Villa – new farm – it was known colloquially thereafter as Newton. "Abbot" was only added to the official name when the station was given that title by the GWR to distinguish it from other towns of the same name on the system.

Brunel's broad-gauge trains of the South Devon Railway reached Newton Abbot in 1846, initiating the change from an agricultural market town dealing in wool and leather into an industrial centre. The railway's locomotive works and depot were situated here and greatly expanded when the GWR took over the South Devon in 1876. When broad gauge ended in 1892, the works here had the heavy task of converting all the surviving engines to standard gauge. Other industries moved in and the population, 1,623 in 1801, soared to 12,518 in 1901.

Serving branch lines to Moretonhamsptead and Paignton, Newton Abbot thrived well into the 20th century, with loco-motives always under repair in the works – I remember seeing Halls and pannier tanks in the 1950s – but the early demise of steam on the Western Regions in 1962 put an end to its fortunes. The station once had nine platforms, now reduced to three, and trains can bypass it altogether.

The works outshopped its last engine, 4566, in 1966. The sheds were "dieselised" but also closed four years later. This unusual railway town may have been on its uppers after the decline of the railway, but it hardly deserves its 2023 description in the *Daily Telegraph* as "one of the ugliest towns in Britain". Newton Abbot has a local history museum that tells the railway story, and the heritage South Devon Railway at nearby Totnes keeps GWR history alive.

My second example is Melton Constable, deep in the heart of agricultural East Anglia. It appears in the *Domesday Book* of 1086 as Maeltuna and practically nowhere else until the arrival of the railway in the 1880s, when a new town was built at the junction of four lines linking Norfolk and the towns in the county to the Midlands. An imposing station with an 800ft platform was built, with a special waiting room for the local squire, Lord Hastings. The Midland and Great Northern Railway sited here its main workshops – sometimes titled with fierce, but exaggerated, local pride "the Crewe of North Norfolk". A total of 19 locomotives were constructed here. Crewe managed 7,000.

But the population soared from 100 or so to 1,157 in 1911, regarded as its heyday. The boom did not last long. When the M&GN was absorbed into the LNER in the grouping of 1923, the works were run down and shut entirely in 1934. British Railways began withdrawing services as early as 1959 and Dr Beeching delivered the coup de grace in 1964 when the entire system was closed. The station was demolished in 1971 and the works are now factory buildings. It is hard to believe that there was ever a railway here. Melton Constable's population is now a little over 600 – half the size of the historic railway town.

There is talk of the railway returning as part of a North Orbital Railway, between Holt and Fakenham via Melton Constable, creating a possible circular route through mid-Norfolk linking to an operational main line. The project has the backing of the Campaign for Better Transport, and Norfolk County Council has protected the route from development prejudicial to reopening the line. But it is only Priority 2, and it is hard to see a revival of the "Muddle and Go Nowhere" line in the near future. Renamed the "Missed and Greatly Needed" after closure, it was clearly more missed than needed, an all-too-common lament for towns that have lost their railway.

CHAPTER 11

HOKEY-COKEY PRIVATISATION

FROM WHEN IT was first proposed in 1844 by William Ewart Gladstone, President of the Board of Trade and later Prime Minister, it took 140 years for nationalisation of the railway to happen. In between, there were two periods of state control during wartime and an enforced amalgamation of the private companies.

After Labour came to post-war power pledging to public ownership, the railway was nationalised in 1948. It took another 48 years for the Tories to re-privatise the system and a further 28 years for Sir Keir Starmer's Labour government to de-privatise again, or renationalise, depending on your point of view. It seems that whoever is in Downing Street, the politicians and the railway barons just cannot keep their hands off each other. It is a fatal attraction.

For the hapless passenger, this hokey-cokey ritual dance is of limited appeal. He or she just wants the train to come on time, and at reasonable expense. Unfortunately, neither form of ownership offers that guarantee. It is the holy grail of the railway and nobody has ever found it, not least the ceaseless procession of Transport Secretaries charged with the search. Their job is a political graveyard, not a stepping stone to Number Ten. Name one who has made the leap. I can't.

Not even Margaret Thatcher truly believed that the private sector would make a better shot of the job. As the girl from Grantham, a major LNER centre on the East Coast Main Line, she instinctively understood that only railwaymen (and women) can run the railway. All the rest, as Arthur Seaton says in *Saturday Night and Sunday Morning*, is propaganda.

Mrs T reportedly slapped down her hard-line Tory Transport Minister Nicholas Ridley when he suggested flogging off the trains, just as they had water, gas, electricity, phones and the rest of the family silver. "No, no, no!" she expostulated, then – perhaps thinking it was something to do with the European Union – "Railway privatisation will be the Waterloo of this government! Please never mention the railways to me again!"

This must have been in the mid-eighties, and the privateers had to wait a few more years before the lady who wasn't for turning changed her mind. Shortly before her resignation in November 1990, Mrs T accepted the arguments for privatisation from her Chancellor – and heir apparent – John Major. And when he took office, he steamed ahead. Full speed ahead for the buffers!

His solution was not popular capitalism like the previous

sell-offs. There was no "tell Sid" campaign to persuade voters to buy shares in the new railway. There were none to buy. The industry was carved up into dozens of train operating companies, under heavily-subsidised franchises "won" in competition by the lowest bidder. The infrastructure – track, the 2,500 stations, depots and the rest was owned by a separate company, Railtrack. And the trains were sold for £1.8billion to a third group of leasing companies – the Roscos (rolling stock companies), at a very substantial loss to the taxpayer. If ever there was a recipe for chaos, this was it. And it duly came, about the only thing that was on time.

Having, against all expectations, won the 1992 election with rail sell-off in their manifesto, the Conservatives pressed ahead with the Railways Act the following year. They were anxious that the benefits of privatisation should be apparent to the travelling public before they were compelled to face the voters again, at the latest in the summer of 1997. Authority was vested in a plethora of bureaucratic bodies, including the Office of Rail Passenger Franchising and the Office of Rail & Road. As Christian Wolmar writes, "The whole Byzantine system was a massive experiment – untried elsewhere before or since – foisted on a reasonably well-functioning industry for ideological reasons."

Tory MP Robert Adley called it "a poll tax on wheels", referring to the ill-starred Community Charge on the number of adults in each household that had to be abandoned. Unfortunately, privatisation was not. Railtrack was the first to go in 1994, followed by the franchises for companies like Richard Branson's Virgin Rail. Some operators soon found that their cut-price bids, based on expected savings and rising passenger numbers, were overly-optimistic. It was harder to run a railway than they thought, even with one hand deep into the taxpayer's pocket. The losses mounted, as did the subsidies, reaching £5billion within ten years. A disastrous string of accidents: Southall, 1997, seven dead; Ladbroke Grove, 1999, 31 dead; Hatfield, 2000, four dead and Potters Bar in the same year, seven dead, was blamed, at least in part on the post-privatisation structure of the industry, which relied on a complex web of contractors and subcontractors to keep the system going. Railtrack went bust and was replaced by a not-for-profit company, Network Rail, effectively renationalisation, by Labour Transport Secretary, Alistair Darling, in 2002.

Labour had come to power in 1997, with a vague pledge to deliver a "publicly accountable, publicly owned" railway, but, like many of Tony Blair's promises, it came to virtually nothing. Perhaps realising that things could not go on forever as they were, Darling did, however, institute a rail review in 2003. That resulted in yet another Railway Act of 2005, abolishing the Strategic Rail Authority that Labour had set up only five years previously, but essentially left things as they were. There were some bright signs. Passenger numbers were rising sharply, safety had improved after maintenance was taken back in-house and timekeeping was almost at pre-privatisation levels.

Next, under various Tory premiers, the government promised major improvements in the railway. HS2, the high-speed link from London to Birmingham, then Crewe,

Manchester, and Liverpool with the ultimate aim of reaching Scotland and the East Midland, served by an extension to Leeds and Sheffield. There would be a new East-West link across the Pennines, joining Liverpool and Hull in what some called HS3. Construction of the first phase went ahead, despite opposition from communities where the tracks would go and their Conservative MPs. But as the years went by, and costs soared into the financial stratosphere, bit by bit the grandiose scheme was scaled back until only the Birmingham route survived. HS3 was reduced to an "upgrade" Integrated Rail plan, with electric trains from Manchester to the great cities of Yorkshire halting at the Pennines. And so the industry soldiered on, with one operating company after another handing back their franchise to the Department for Transport after the operators failed to make the kind of money they had in mind.

Their ham-fisted efforts were not helped by making redundant too many experienced (and relatively well-paid) drivers, putting their hard-nosed union, Aslef, in a seller's market, as the operating companies competed for scarce skills. The devolved governments took their trains back into public ownership: Transport for Wales in 2021, and Scotrail the following year. Over the next few years, four major operators, including LNER and Transpennine, were taken over by the government and run by the Department of Transport's Operator of Last Resort, an office clearly titled by a civil servant with a sense of irony.

The Labour Party, under Sir Keir Starmer, promised to end the chaos once and for all. Shortly before the 2024 general election, he declared, "We've tried privatisation for two or three decades and it's a complete mess. Everybody who travels on the trains has been affected by the cancellations and delays. We have to pick that up and fix it, we can't walk around that problem any more."

His solution was a gradual renationalisation – though the word was not used, for fear of upsetting floating voters – as the passenger franchises came up for review. Another new authority – Great British Rail (after the success of a TV bakery show by that name, everything had to be Great British) – would take over the strategic running of the industry. But it would not be complete nationalisation. Rail freight would remain in private hands, where it had been a success. So, too, would the highly-profitable train leasing companies – the Roscos, which would cost the earth to buy back. As would some privately-financed "open access" operators such as Hull Trains and Lumo. Yet another new body, the Passenger Standards Authority, would "mercilessly" hold the GBR to account.

Naturally, what was left of the private rail sector opposed Labour's plans. While agreeing there was a need for "radical change", Rail Partners claimed that renationalisation would push up costs, lead to higher subsidies and the taxpayer would lose out – as if this was not precisely the record of privatisation. The privateers urged a publicly-owned GBR (as had also been promised by the Tories, in draft legislation that fell by the wayside) with private companies running the trains: a "best-of- both-worlds" solution. Given the history of this long-running Great Hokey-Cokey Privatisation

Show, it was always possible for the politicians to come up with the worst of both worlds.

Starmer certainly made a good start down that road, his Transport Secretary, Louise Haigh, resigning only four months after Labour took office in July 2024. Her offence was to have pleaded guilty to a fraud charge a decade earlier, telling police in 2013 that she had lost her work phone in a mugging, but later found it had not been taken. She was given a conditional discharge by magistrates, but the disclosure compelled her to resign, saying that her appointment as the youngest ever female Cabinet minister at the age of 37 was "one of the proudest achievements of my life", which is just as well as she had not had the chance to do anything in the job. Ms Haigh had already been rebuked by Starmer for calling P&O Ferries, a "rogue operator" after its owner, United Arab Emirates-based DP World, threatened to pull out of a £1billion investment in the UK. P&O had previously hire-and-fired its ferry workers.

Heidi Alexander MP, Labour victor in the former railway town constituency of Swindon South, took over the department and its travails. She is no stranger to controversy, having quit Jeremy Corbyn's "entirely dysfunctional, inept, unprofessional, shoddy" Shadow Cabinet in 2016. She has the dubious honour of being the 40th holder of that office, under various titles, since 1945, a collection of parliamentary duds and forgotten aspirants to premiership. None of them lasted more than a few years, Ms Haigh did not set the record for the shortest time in office. That dubious palm goes to Tory MP Anne-Marie Trevelyan, who managed only seven weeks – but that was in the non-government of Liz Truss of 2022.

This inglorious procession of second-raters, none with any identifiable experience or even knowledge of the railway, has vitiated the future of an industry on which the country's economy, and the travelling public, relies. Since the surge in road transport and nationalisation of rail in post-war years, the Department of Transport has been consistently pro-road and, if not anti-rail, then unfriendly towards the railway, regarding it as an expensive, troublesome nuisance – a state of affairs to which it has conspicuously contributed.

Ms Alexander, a former deputy mayor of London under Sadiq Khan, may be the first to remedy that. She served as deputy chair of Transport for London and on the board of Crossrail, tackling delays to the opening of this hugely-efficient and successful new East-West commuter link through London from Reading and Heathrow Airport to Shenfield in Essex. Named the Elizabeth Line after the late Queen, it was completed in 2022 at a cost of £18.8billion, but now has annual revenues exceeding £1billion, with 700,000 passengers journeys on weekdays. If she brings to the job the vigour shown in her determination to unlock the Crossrail construction "pause" created by the Covid pandemic, perhaps there is hope for the 21st century railway yet.

CHAPTER 12

TRAINSPOTTING

IN THE 1950S, no self-respecting main line station was complete without a motley crew of schoolboys congregating excitedly at the end of the platform, clutching notebook and pen and eagerly scanning the signals for the next train.

Trainspotters were not then the anti-heroes of Irvine Welsh's dystopian drugs novel but avid collectors of engine numbers. Had you looked carefully, you might have seen me among them.

We lounged about on the heavy wooden porters' barrows, or squatted on the stone platform (never with legs dangling over the edge, that would have prompted immediate banishment from the station: staff kept a kindly eye on us). If it rained, we took cover under the canopy, causing a bit of a nuisance to "proper" travellers and annoying the porters looking for customers (and tips).

It's fair to say "proper" because though some may have bought tickets and travelled many miles by train, for others this was just their local station. Children did actually live in Crewe, and they must have been among the dozens who congregated on the long pedestrian bridge high above the northern end of the platforms, with a bird's eye view of the trains, and the comings and goings from North Shed. It was a spotter's paradise and I must have been in that arcadia at least ten times.

Stations were not as secure as they are today, with automated ticket machines. There was usually only one ticket inspector and you could easily dodge by him, or he might turn a blind eye. But if necessary, you could buy a platform ticket, costing one penny (less than half a "p") and valid for an hour, really meant for people seeing off friends and family. The tickets were so constructed with numbers round the edges so they could be clipped for the date and the hour, knowing when your allotted time was up. I have never known this discipline to be enforced.

Rucksacks of the kind universally worn today were unheard of. We either squeezed our pop (more likely, water) and sandwiches into our pockets or stuffed them in a stout, canvas ex-army gas bag, fastened with brass buttons and carried over the shoulder. This was a key part of the spotter's uniform, along with the anorak then coming into fashion. The big must avoid: getting jam from the sandwiches on your numbers book, or worse still on your Ian Allan ABC book of locomotives.

These were the spotter's bible, with the numbers and names of all BR locomotives, together with key facts and statistics. They were the brainchild of Ian Allan, who worked in the publicity department of Southern Railway during

World War Two. Harried by constant requests about his company's rolling stock, he collected the information into a book in 1942. This first volume, *ABC of Southern Locomotives*, was an instant success, feeding the schoolboy fascination with engines, particularly steam.

It was followed by further books on different companies, and then by combined volumes incorporating them all in one cover. The "combine" was a treasured possession, supplemented by the *Locoshed Book*, which carried only numbers and the details of every home shed of every locomotive in Britain. That was for serious spotters, who "bunked" (illegally entered) the motive power depots in search of "cops (first-time sightings)".

After the war, Ian Allan Publishing grew hugely and diversified into other fields including military and civil aviation, but ceased interest in the trainspotting business and finally closed in 2020. Classic editions of the ABC books have been reprinted, and early, pristine copies command serious prices in the market for railwayana.

Trainspotting got teenage boys into a bigger world than the town where they lived, taught resilience, geography and the skills of travelling to distant parts – a sense of adventure. The hobby has few advocates today, but Portsmouth-born Michael Harvey, in his memoir *Forget The Anorak: What Trainspotting was really like*, defends his days out with the ABC and dismissed the risks involved in night-time visits to depots. "We were oblivious to those kinds of danger, probably because we were teenagers, and teenagers in those days had little fear of anything or anyone."

"What a grand feeling," he recalls, "and sense of achievement as we totalled up our cops on the homeward bound train, knowing that we had successfully 'bunked' engine sheds crammed tight with living steam and eluded the foreman, and taken photographs that in later years would become gems of nostalgia."

Pop music legend Pete Waterman's love affair with the steam railway goes back into his infancy. "I was born on the coldest January day of the Big Freeze in 1947 and almost the first thing I heard was a Super D pulling a load of coal waggons on the line from Coventry colliery outside my window," he told me in an interview for the *Daily Mirror* in 2015.

The Super Ds were the workhorses of the old London and North Western Railway built before World War One, but some survived into nationalisation and the 1960s. "When you live in a council house and one of these things goes past your door, it's your first encounter with beauty."

Pete was an avid trainspotter on Leamington Spa station, watching the GWR King and Castle class locomotives roaring their way through to Birmingham. When he left school in 1963, he got a job at Wolverhampton Stafford Road station, where these magnificent engines were serviced.

"They said, 'You look like a fireman' and that was it. I spent three weeks going up and down the sidings learning how to fire and then I was on the footplate with a shovel. And I was crap!"

His first time out on the job was a turn firing an old GWR 28XX class from Wolverhampton Oxley to Severn Tunnel

Junction in South Wales. "I was a young lad with only a month on the footplate, trying to keep an engine in steam and it was impossible. I found out later that the engine was knackered anyway.

"I was very lucky. The driver, I thought he was an old boy – he was probably about 50 – he told me 'take your time'. He taught me a lot." Steam had not long to live on the BR Western Region, and Pete was on the footplate for only 14 months before going to work at GEC's Coventry plant, where he became a TGWU works convenor.

But his passion for steam engines never left him. He spent £700,000 on restoring one of his beloved Super Ds: 49395, built at Crewe in June 1921 and withdrawn from service in 1959. The locomotive was saved for the national collection, but languished at sites all over the country before Pete stepped in with an offer to make her good as new. 49395 operated on heritage lines, and is now on show at the NRM Shildon.

The story does not end there. In April 2015, he sold his unrivalled collection of magnificent scale-replica engine models for £627,000 to set up a heritage restoration works in Crewe, cradle of his life's passion, the LNWR. There, his Pete Waterman Railway Trust is training young people to become the next generation of locomotive engineers. *A Train Is For Life* is the title of one of his books. And that's true for so many.

Spotters all had their favourite classes of engines, mostly born from the ones they saw daily and got to know best. Even the dirty, smokey old goods engines had their familiar charm. Home also mattered, particularly when you were only a nipper. Unless your parents lived near a main line station, or a junction bringing several routes together, chances were you saw little but locally-based locomotives, going about their regular passenger or goods business. Humble shunters, branch-line tank locos. Not the LNER A4 "Streaks" or the LMS "Coronation Scots".

As you got older, and more ambitious, boys (they were virtually all lads) would travel to the nearest big city, or to regional centres like Crewe, where the variety was greater and there were motive power depots [engine sheds] to sneak around. There were engines in much greater numbers, sometimes upwards of 100.

Admission was officially forbidden, with 'Trespassers Will Be Prosecuted' notices at the entrance. For years, I thought this was something to do with the Lord's Prayer. It was possible to gain an official permit to enter these hallowed halls, with turntables, coaling plants and all the paraphernalia of steam. But that meant writing to the BR public relations office, with the endorsement of a responsible adult.

MPDs were dangerous places – the men who worked there were sometimes injured or even killed. Long lines of engines stood over deep ash pits, tempting the unwise to jump across the rails.

But the lure of the shed was greater than the sense of caution for a spotter in search of a "cop". I admit to trespassing into my home shed of Normanton, West Yorkshire [coded 20D in the LMS era, and 55E in the North-Eastern Region of BR]. It was easily reached by climbing a tip at the open end, or by bending double to sneak past the foreman's office at the entrance used by drivers and firemen. What fun!

More seriously, I cracked the massive two-turntable Leeds-Holbeck [20A in LMS days, 55A in BR] by climbing up a wall and over a fence, or even more daringly, through an open window at the back, a feat that required all my body-bending skills. Risky, I know, looking back. But it was part of the game, like dodging the foreman.

I even sneaked into the famous "Plant" in Doncaster by a form of limbo-dancing through an open window in the paintshop, much to the awe of fellow-spotters gathered below, too frit to follow my dare. It being a Saturday, with no workmen on site, I wandered through the erecting shop, the Crimpsall where Flying Scotsman was built and along the scrap lines where condemned engines awaited the torch.

Their names eventually disappeared from our ABC books, so not as valuable a "cop" as the Sandringham or Antelope classes from distant Eastern Region depots. I was never challenged, but I don't remember how I got out – certainly not the way I got in.

Footplatemen (they were all men in those days), cleaners and other shed staff – also all men – rarely bothered us, though I was once cornered by a fireman at Doncaster Carr MPD [36A throughout] who took me on the footplate of unique 60700 which he claimed to have fired to Kings Cross. He confiscated my notebook of numbers, which I thought – and still think – was a mean trick. It was of no use to him.

And not, in the end, of much use to me. In my mid-teens, I joined the Wakefield Railfans' Club, which met in the basement of a chapel in the city. This was to be my downfall as a trainspotter, a fate that I shared with most of my fellow number-crunchers: girls. One girl, in particular, the young sister of a city railfan, a devotee of the GWR, an arcane interest since we were at least 80 miles from the nearest Western locomotives.

Suffice to say, I went home to Railway Terrace one night in 1961 after a club meeting and threw all my train books, with collections of numbers from all over the country, gathered with painstaking effort (and probably worth a tidy sum now) into the bin. My mother gave me a knowing look, but said nothing.

She was right. I had discovered the opposite sex, as it was then called. I am still discovering it after more than 60 years of marriage to one of the girls who turned my head that summer.

But now, in a kind of second childhood, I can explore what's left of the great British steam locomotive fleet – I was never interested in diesels – on more than 100 heritage lines in far-flung places. Give or take – some were scrapped after being "saved" - there are 375 surviving engines from Aviemore to Bodmin.

The consolation of second time round is that you have more money, more opportunity and more patience. So it's down to the station with the anorak, the rucksack with sandwiches and pop (er, more likely to be pinot grigio now) for a nostalgic journey. It's as easy as ABC.

My enthusiasm for steam locomotives almost got me jailed as spy in communist Yugosalvia during 1981. Touring provincial Serbia with a newspaper pal, I spotted a line of engines at the sheds in Kraljevo, and persuaded him to join me in a foray. We had not gone many yards before being detained by a very young, excitable soldier with a very old rifle, who called the police.

Accused of spying, we were taken to the local cop shop, and interviewed separately, partly in German, the only other language they knew. After a couple of hours, they were satisfied that we were not agents of MI6 – my puny Kodak instamatic camera being evidence of amateurism rather than espionage – and let us go. We celebrated with a glass of slivovic [plum brandy] at the railway station. We were lucky: plane spotters had been jailed for years, but we convinced them there was nothing hush-hush about trains-potting, just another strange English eccentricity.

And a new generation is taking to the platform, not physi-cally but digitally. More than two million people follow social media personality Francis Bourgeois (real name Luke Magnus Nicolson), whose light-hearted and humorous trainspotting videos on TikTok and Instagram have taken the hobby into a new dimension. The 25 year-old now has his own company, acts as a brand ambassador for GB Rail-Freight, promotes other companies, appears on TV and has authored an autobiographical novel, *The Trainspotter's Notebook*.

It's all a far cry from my teenage shed-bashing days of the 1960s. I wonder what he would have made of the steam age, which ended more than 30 years before the businessman-spotter was born? What, indeed, would the pioneering

Quakers of Darlington have made of him? I rather think they would admire his enterprise, but it's not my cup of British Railways tea.

CHAPTER 13

TRAINS, BOATS, HOTELS AND PLANES

DINERS, DRINKERS AND residents in Manchester's luxury Midland hotel attending nearby party political conferences rarely, if ever, give a thought to their elegant surroundings. I know, because I have dined and drank (but mostly the latter, except for execrable canapes) as a political journalist staying there while covering Labour and Tory gatherings over many years. Nor do many of the delegates – representatives, as the Conservatives insist, because they are not delegated by the blue-rinse brigade back home – realise that the grandiose Manchester Central complex in which they expel so much hot air was once filled with smoke and steam from engines bringing trains under its majestic roof, when it was built as a railway station by the Cheshire Lines Committee in 1880.

The trains have gone, as in so many other places, but the hotel lives on, a testament to the vision of the railway builders of the 19th century. They quickly realised that where there are travellers, there must needs be beds and food and drink. And where better than where the trains are? Almost the first off the mark was the Midland Railway, with a hotel at the bottom of my street in Normanton, where Queen Victoria and her consort Albert once dined while changing trains. The building is still there, but no monarch is likely to pay a visit. In my youth, what remained in commercial use was a pub known as "the Taps", later transmogrified into the Flying Scotsman with much railway regalia on the walls and ceiling. Alas, no more, much like the station itself.

The London and Birmingham was also among the first out of the trap, opening hotels at Euston and Curzon Street when the main line began operating in 1839. By 1854, 30 more had opened across the country, and by 1913 this figure had swelled to 93. And what hotels! They were built to high standards, with imposing architecture and what would now be called hi-spec accommodation. They gave employment to thousands of staff, and promoted the company brand. Architectural historian Nikolaus Pevsner calls the earliest examples "the most interesting hotels of the 1840s".

They became a byword for comfort and convenience, often being connected to the station they served for ease of transfer. Travellers need not even see the town or city where they broke their journey.

Sometimes, like the North British in Edinburgh, the St Pancras in London and the Royal Station in York, they were landmarks on the skyline. Others, like the Adelphi in Liverpool, the favourite of Labour premier Harold Wilson, accommodated royalty and statesmen. They were also places where businessmen did deals, and couples discreetly dated. They were a busy world in microcosm. I have stayed in at least a dozen of them, for work and latterly for pleasure, most recently the Old Station, now the Royal Highland, in Inverness. The sense of history is almost tangible, though in the latter days of public ownership it was alleged that standards had fallen. One critic complained that "the enjoyment of dinner must not get in the way of staff preparations for breakfast".

At nationalisation in 1948, 55 hotels were handed over, including 26 by the LNER, boasting 2,429 bedrooms, 17 by the LNER (2,328), four from the GWR (336) and two from the Southern (224). Between them they accounted for over a million guests. They remained in public ownership, under various different guises, for 30 years, withstanding even the scalpel of Dr Beeching. The election of Margaret Thatcher put paid to that. The hotels were originally built by the private sector and that is where they must return. Top people at British Transport Hotels offered a management buy-out, but the Tories insisted they be sold singly, even though the taxpayer was ripped off with lower sales prices.

Some hotels today form part of a great international chain. The Caledonian, at the end of Princes Street, that once served the now-defunct station of that name, is now the Waldorf-Astoria Edinburgh, the Great Eastern is the Andaz Liverpool Street, the North British has become the Balmoral and the Royal Station York as the rather more mundane The Principal. Some, like the Zetland in Saltburn by Sea, the Sandringham, Hunstanton and the Felix, Felixstowe have succumbed to changing conditions. The Turnberry, Ayr, is now owned by US president Donald Trump, and is a golf resort. Trump Towers-by-the-Green offers rooms averaging £500 a night, hopefully with full Scottish breakfast, also known as a heart attack on a stick.

The railway companies didn't just put you up for the night. They fed you on your journey, in what were always known, rather quaintly, as refreshment rooms, whether they were simple tea shops or boozy bars. I have lost count of the number I have enjoyed (patronised would be too elevated a term) and I hope to enjoy many more. The scene has changed utterly over the years. Where you once had scones and coffee, you can now slake your thirst at a Wetherspoons, in the grand hall at Leeds, or in The Draughtsman Alehouse on Platform 3 on Doncaster station.

So popular have these bars become that railway Ale Trails are greatly in vogue for a certain class of imbiber: the real ale enthusiast. They are amply catered for on the Trans-Pennine Trail at stations like Huddersfield, which has not just one but two pubs: the Head of Steam and the Kings Head, while nearby Dewsbury offers the West Riding Refreshment Rooms and over the border in Lancashire is the renowned Station Buffer at Staybridge, coal fire museum of historic rail artefacts and eight beers on the handpump, including Timothy Taylor's Landlord, a brew not to be taken lightly.

The craze for beer safaris is so great that Bob Barton, a former press officer with VisitBritain, has published a coffee-table sized guide book to the licensed treasures of the railway, describing 74 in detail (plus another seven in brief). And, hero of the railway that he is, he visited them all. Sometimes, by bike. You can't always rely on the train, especially on heritage railways that might only run in the tourist season.

Hotels and watering the thirst traveller were by no means the limit of the pioneering companies' imagination. Where there was traffic, there was profit. The first public railway ran to a quayside, and railways diversified into docks, the pre-grouping companies taking 28 facilities into the LMS, ranging from piers for steamers on the Clyde to the ports of Holyhead, Heysham and Stranraer for the Irish trade. During World War Two, they had to adapt to new traffic and ways of handling. Fish harbours and banana ports handled coal, and they all found themselves under military control. They were also prime targets for the Luftwaffe, whose relentless bombing during the Blitz made docks the most-hit places outside London.

The Southern Railway inherited a clutch of facilities in the Thames estuary and along the south coast, from Whitstable to Queenborough down to Dover, Folkestone, Newhaven, Plymouth and even Padstow. The jewel in the crown was the key port of Southampton, which as we have seen was critically important in the mass movement of men and material during both world wars, as well as being the main destination for Trans-Atlantic liners, overtaking Liverpool.

The Great Western, being largely an inland system, had fewer docks, mainly at Plymouth and Bristol and in south Wales for the export of coal. But it had one unique offering: the train through the streets of Weymouth to the ferry for the Channel Islands. This short line, built by the GWR in 1865, ran from the main station to the harbourside and began ferrying holidaymakers to the ships in 1889. I have vivid memories as a boy of about 12 on a family holiday to Jersey, trundling slowly through the streets of the resort behind a tiny pannier to the pier head. A man with a red flag walked ahead of the train, warning pedestrians to get out of the way. I also recall a bell on the engine buffer beam, though that may be a detail too far. The line closed for passengers in 1987 and the Office of Rail Regulation approved total closure as recently as 2017. Today there is little trace of this memorable little adventure.

In 1923, the LNER swallowed harbours in 20 ports, including Grimsby, Hartlepool, Hull, Immingham, Middlesbrough, Harwich, Lowestoft, London and outlets on the eastern seaboard of Scotland, principally Aberdeen. Coal was exported and freighted to London from huge staithes like those at Blyth that survive today. Iron ore was imported at Tyne dock for the steelworks at Consett, County Durham, a lucrative trade that continued right into the 1980s. The company also ran New Holland Pier, where travellers took the ferry from Lincolnshire to Hull, a journey made redundant by the opening of the Humber Bridge in 1981, at the time the longest (2.2 km) single-span bridge in the world. I have distant memories of taking this ferry with my father, an inveterate collector of unusual journeys by rail.

With docks naturally come ships, and here again the enterprising Victorians were busy. For example: the Manchester, Sheffield and Lincolnshire Railway commissioned

1,000-tonne sister ships SS Lutterworth and SS Nottingham in 1891, for the Grimsby-Hamburg run. In the years before nationalisation, the LNER had operated 41 ships. And at the outbreak of World War Two, the LMS commanded a fleet of 49 sea-going ships, sailing out of ports on the Clyde, the North Sea and the Irish Sea and the inland port of Goole, Yorkshire, plus pleasure steamers on the English Lakes and Scottish lochs.

"Steamships, motor vessels, paddlers, turbines – they were of all shapes and sizes, aged and pedigree, ranging from little 30-ton launches to the large and luxurious steamers of the Irish Mail route," boasted the LMS war historian. "Had it been possible for them all to be present at a mercantile review they would have numbered over 100 strong with a gross tonnage of well over 80,000." They saw valiant war action as troopships and vital cargo carriers. No fewer than eight vessels on active service were lost, including the almost-new Princess Victoria, built in 1939 for the Stranraer-Larne route.

Accordingly, at nationalisation, the railway inherited a mixed bag of ships, some of which were old and scrapped. But the main routes were eventually constituted in 1970 as one company, Sealink, operating services to Ireland, the Channel Isles and continental Europe, in a consortium with French, Dutch and Belgian shipping companies. The Tories sold Sealink UK to Sea Containers for £66million, which then passed to the ownership of Stena Line in 1991, as now rebranded.

Last, and counter-intuitively, the railway ran airlines. This might not seem unusual in an age when Richard Branson's Virgin Airlines also owned a big train company operating on the West Coast Main Line for more than 20 years, but it was revolutionary at the time. Railway Air Services, a purely domestic airline, was formed in 1934 by the Big Four companies to serve London and the major provincial cities from a base in Croydon. It was taken over by the government at the outbreak of World War Two, and restricted to the London-Glasgow-Belfast route for mail and "priority" passengers. After the war, services were resumed, with Ansons, ex-RAF Dakotas and even ex-Luftwaffe craft, but along with all the other assets, the airline passed into public ownership with British European Airways in 1947. During its brief life, eight crash incidents occurred, with the loss of an estimated 17 lives.

No-one could say that the railway had an inward-looking, blinkered mentality. It always embraced the new, even operating Hovercraft across the Channel and to the Isle of Wight. The history of the industry is one of innovation, even if it didn't always get things right, as we shall see with the sorry story of the Advanced Passenger Train.

CHAPTER 14

THE LINE THAT REFUSED TO DIE

IT'S 05.50 AND I'm on Platform 2 of Cononley station in North Yorkshire, waiting in clear but chilly weather (5°C) for the direct diesel train to Carlisle. All the electric trains from Leeds to Skipton call here, but this is the only one of its type that does so daily, en route to the border city. Mine not to reason why, just to be thankful I can catch it, for this is my annual safari to Scotland.

The path to the platform takes me past a showcase of art and nature by the Friends of Cononley Station. A large colour portrait of an otter bids me welcome. Half a dozen tiled flower mosaics set into gravel adorn a selection of big wooden flower barrels and plastic tubs. This is daffodil time and they are at their best, nodding their heads in a thin breeze. Nobody asks local people to do this. They do it for the love of the railway and the life it brings to the village. The station was opened in 1847, but closed with the Beeching Cuts in 1965 and only reopened in 1988 by British Rail with help from local councils and the Rural Development Commission. It now serves an active commuter community to Leeds and Bradford.

The forecast is for warmth, but there's a nip in the air before dawn. In the shelter on Platform 1, a lady waiting for the train to Leeds strokes the station cat on her lap.

"What's its name?" I call across the tracks.

"Una," she replies. I suppose it might be Oonagh, but that's a question too far. Her train is on time and off she goes, and so does the cat, a big brown creature I see frequently.

My route today takes me over the famous – some would say infamous – Settle to Carlisle railway: the line that refused to die, though it had many willing executioners in the sixties and eighties. That it is still going today bears testimony to the extraordinary battle waged by supporters and enthusiasts, plus some nifty footwork by Conservative politicians such as the great railway TV presenter himself, Michael Portillo, eager to take the credit. This was always the Cinderella of the iron road to Scotland, and it shares the fate of the fairy-tale princess – undreamt of success against the fates.

The Settle and Carlisle (S&C) line, winding its way for 73 miles through the inhospitable terrain of the Pennines, was built at the last gasp of Victorian railway mania. It serves no centres of industry or cities – or even middle-ranking towns – and the few that it does often condemn travellers to a long uphill trudge to get to the trains.

My journey goes over the highest point of the railway in England – the 1,169 ft summit at Ais Gill, strides across wild moorland on massive stone bridges like the incomparable

440-yard, 24-arch Ribblehead Viaduct, plunges through deep gritstone tunnels like 2,629 Blea Moor and is subject to the wildest winter weather, kept at bay by palisades of snow constructed from old railway sleepers. It is the permanent way engineer's nightmare.

This Spring morning, not so soon after the clocks have gone forward, I will see the sun rise through the carriage window. Faint rays are already appearing over Skipton Moor, dark in sharp outline. Homer's "child of morning, rosy-fingered dawn" in the epic story of Odysseus appears as we leave the station. This is no odyssey, but it's the best we can do.

At Skipton, known as "the Gateway to the Yorkshire Dales" we pick up a few passengers at this classic, grade 11-listed Midland Railway station, rebuilt in 1876. It has seen more important days. The branch line to Grassington in Wharfedale was closed to passengers as early as 1930, and the connecting route to Ilkley in 1965, as was the line to Colne in Lancashire in 1970. Enthusiasts are trying to reopen the latter, but it seems a bootless quest.

At our first stop, Gargrave, nobody gets on and nobody gets off, like a scene from Edward Thomas's poem *Adlestrop*. The closed station of Bell Busk, in the past touted as the stop for the tourist attraction of Malham, flashes by. It's still dark and silent at Hellifield, once a major junction with an engine shed where locomotives with snowploughs were based.

Settle station, 236 miles from London and 72 to Carlisle, is a handsome example of "Derby Gothic" architecture with cream-coloured, decorated bargeboards. Its low stone buildings emerge in the half-light, while in the yard an iconic water-tower-turned-residence that starred in the TV show *The Restoration Man* is just visible. On the up (to London, always) platform the wooden seatbacks say 'Settle Up', and across the tracks, it is Settle Down'. A nice touch.

From here, we begin what the footplatemen knew as The Long Drag, climbing for the next 16 miles, at 1 in 100. Between here and Dent, the highest main line station in England, the line climbs 200 feet alone.

At our next stop, Horton in Ribblesdale, we are at the border of rivers: from here, the Ribble debouches into the Irish Sea, behind us, the Aire into the North Sea. And in the middle distance to the right, rearing up in the lightning sky, is Pen-y-Ghent ("hill of the winds") 2,277 ft. It is finally dawn at 06.36 by the time we get to Ribblehead, the station of the famous viaduct. There is only one other person in my carriage, and nobody comes to check my ticket. I enjoy a ghostly communion through the window with the bleak moorland, as our diesel powers slowly across the great 24-arch structure high above Batty Moss, a bog pitted with mysterious sink-holes from the time of navvy settlements. Of course, the best view is from below, but there are no cars parked in the best photo-spots today.

To our left is the 2,415ft massif of Whernside ("hill where millstones are found") and Ingleborough, 2,372ft, which I have once climbed but am unlikely ever to do so again. Leaving Ribblehead, our train dives into Blea Moor tunnel, a mile and a half long, and a sore trial to footplatemen battling up the Long Drag in steam days. This morning, we burst through in a couple of minutes, emerging at the head of beautiful Dentdale, home of the "furious knitters" of folklore.

The locals needed strong legs as well as deft fingers, because Dent station, the highest in England at 1,139ft, is five miles and 600ft above the village. It is a lonely spot, with the ruins of snow fences leaning crazily on fields above the track. And it gets lonelier at Garsdale, once a junction with the LNER line to Hawes and thence to the charms of Wensleydale. Here there was a turntable in the steam days and legend has it that one locomotive was caught in a gale and whizzed round and round until some bright spark halted the whirligig with a bucket of sand on the mechanism. The line closed in 1959, but a heritage railway now serves Wensleydale and the Little White bus goes from Garsdale station yard to Hawes.

On, relentless on, for a few more miles, over Dandy Mire, through Moorcock tunnel and over Lunds viaduct, where a bad smash caused by signal error on Christmas Eve 1910 cost the lives of 12 people. Finally we reach the summit at Ais Gill, 1,169ft above sea level, the roof of England's railway. By now, just before 07.00, the sun is a blazing orange ball of fire in a clear blue sky. Entering the lovely valley of the river Eden, it's downhill all the way from here: literally and figuratively. The Cumbrian scenery has little of the harsh hill and moorland spectacle on the Yorkshire side. Even Cross Fell, the highest summit of the north Pennines and the highest in England outside the Lake District at 2,930ft fails to impress with its flat top.

Swooping down the 1 in 100 decline, we reach Kirkby Stephen station, though not the town itself, because it's two miles away and 200ft down a steep hill. The station buildings are now self-catering accommodation, attractively situated but a long way from the shops. We rattle on through Crosby Garrett, station closed in 1956, and Long Marton, New Biggin and Culgaith, all closed in 1970. To be fair, they served remote rural villages, and there was little hope of revenue in the age of mass motoring.

At Appleby, once the county seat of Westmorland (as a station sign proudly announces), we are in rich farming country. In the far distance rise the peaks of the Lake District, or so I presume. What else could they be? Passengers bound for work and school in Carlisle begin to fill the empty seats. There is chatter, which is fine, and iPhone braying, which is not. There is no quiet coach on this line, or elsewhere in my recent experience and I fear the concept may now be history. The tech companies have won on the railway, as everywhere else.

The river valley is full of fog, and melting frost glistens in the field as we reach Langwathby and then Lazonby and Kirkoswald, names that sound just as pretty as the villages they serve once again, after being opened in 1986. Little Salkeld, closed in 1970, and now a private home (Oh! What a place to live!) was the scene of yet another fatal accident. On 19 January 1918, a northbound morning Glasgow express ran into a great mass of rock and dirt caused by a landslip, killing seven.

Through the wooded Eden Gorge, we come to Armathwaite, where I peer anxiously through the window for a sight of the castle said to dominate the village but the sun is too strong for me. Past Cotehill, closed as early as 1952, we skirt the site of former Durran Hill engine sheds and join the line for Newcastle at Petteril Bridge Junction, sliding easily into Platform 6 at Carlisle station, on time. The once-empty train now disgorges more than 30 passengers. Formerly named Citadel, and built with great blocks of sandstone like the nearby castle, this is what I call a proper station, very much in Routledge country. My family originated in this part of the world, and the cemetery in the border hill village of Bewcastle north of the city is full of Routledges. Best place for 'em, I have heard said.

Carlisle was once served by seven railway companies, and six of their lines are still operational. A massive stone fireplace with Latin inscriptions adorns the comfortable waiting room, whose walls tell the story of "the line that refused to die" in large poster interpretations.

But it appears that the big refreshment room, with a once-welcoming bar, has not survived. A large metal barrier blocks the door. Still, there is time to savour the feel of a great railway entrepot, with trains arriving and departing every few minutes for Edinburgh, Glasgow – my eventual destination today – the Cumbrian coast (but no longer Silloth), Newcastle, Manchester, Liverpool, Preston and London, and, of course, approximately 16 a day to Leeds via Settle on a line that would never have been built, but for the Victorian vanity of the Midland Railway directors.

They were determined to have their own exclusive route

across the border to Glasgow and Edinburgh in the latter quarter of the 19th century. There was a potential branch-line link with the LNWR at Ingleton, but the Premier Line, jealous of its monopoly of the west coast route, deliberately made that unworkable. The Midland had second thoughts about their £2.3million [almost £300million today] ambitious scheme, and tried to back out but was obliged by Parliament to go ahead.

Construction began in November 1869, not the best time of the year. Midland bosses brought 6,000 mostly Irish navvies – and "wives" and families – to this harsh terrain for the five-year task of conquering the North Pennines. They lived in squalid shanty towns with names like Sebastopol, Inkerman and Jericho. Traces of one, Batty Green, can still be seen below Ribblehead viaduct. Many – it's not known how many, but probably hundreds – died from accidents or disease. Some received an anonymous interment in trackside spoil heaps, marked by makeshift gravestones at the northern end of Blea Moor tunnel. Eighty died of smallpox at Batty Green. A plaque in St Leonard's church, Chapel-le-Dale commemorates the line's victims and a memorial stone in St Mary's Mallerstang, remembers the 25 navvies and their families buried there in unmarked graves.

The line opened to good traffic in August 1875 and to passengers eight months later, by which time costs had soared by 50%. But the line did its job, attracting passengers to crack services like the Thames-Clyde Express, though the journey from London to Glasgow in 1962 took nine hours, as against the west coast's seven hours and 20 minutes. The line was – and remains – an architectural masterpiece. All the stations, originally numbering 19, were built to the same "Derby Gothic" design, a unique ensemble of style.

The first attempt to demolish this "roof over England" came, predictably enough, from Dr Richard Beeching, whose 1963 Report proposed closure of all the stations. Unlike many other routes marked for shutdown, it survived, but services were gradually reduced. All the stations except Settle and Appleby were closed and with the electrification of the West Coast main line, giving much faster times, the Waverley express to Edinburgh was taken off in 1968, and the Thames Clyde in 1975. Only three through trains a day from Nottingham remained and they went in 1982.

Despite protests from local people and the Yorkshire Dales National Park, the end of the line was in sight. Closure of passenger services was announced in 1984, with British Rail blaming the cost, estimated at £6million, of repairing the century-old Ribblehead viaduct. This was disputed by competent engineers and with BR moving goods traffic away from the line, protesters built up a convincing case of "closure by stealth" on the part of BR.

It was a three-pronged approach. First, BR managers offered Cumbria County Council a £7million upgrade to the county's coastal line if they dropped their opposition to closure of the S&C. It was rejected. Second, the cost of Ribblehead restoration was inflated to as much as £10million, an exaggeration professionally rebutted. Third, an experienced new manager, Ron Cotton, was appointed to oversee the shutdown, a process scheduled to take three to six months. This backfired. He set about making the line more viable, by introducing cheap fares, running more trains and reopening eight stations.

The protest gathered pace, with the emergence of a coalition between local councils, business people and Friends of the Settle and Carlisle (FoSCL). Lawyers in London were briefed to challenge the validity of the closure proposals, as out of date and legally flawed. A judicial review was threatened.

Tory Transport Minister Paul Channon was unmoved, stating in 1988 that he was "minded" to go ahead with closure and took the extraordinary step of proposing the sale of the century: the S&C on the market lock, stock and barrel offered to potential private sector investors. "The world's biggest train set" I called it in *The Observer,* with some hyperbole.

At this point, the shutdown momentum stalled. No credible buyer came forward. Intense pressure was put on Conservative MPs and possibly Lord Whitelaw, Prime Minister Thatcher's former deputy who had represented Penrith and the Borders. The threat of a judicial review would delay the process by many months, if not years – and might not go the government's way. Passenger figures and takings rose substantially, partly driven by all the publicity. And in February 1989, an internal Transport Ministry submission to ministers suggested that while the financial case for closure still stood, "either closure or retention could be justified by the facts".

As disclosed in the brilliant sleuth journalism of writer Michael Pearson, whose book on the saga is based on government files released under repeated FoI requests, the final decision was left to Minister of State Michael Portillo. There is no written record of his verdict and he disclaims sole credit, saying only "the result bore my fingerprints".

The denouement came swiftly. On 6 April, Channon wrote to the Prime Minister, pointing out that this difficult case had aroused strong emotion, dragging on for five years. "I have concluded that I should, after all, refuse British Rail consent for closure." He cited new evidence: a significantly less robust financial case, making him less confident of winning a case of judicial review, and the failure of any buyer with sufficient financial backing.

Tellingly, he also observed, "I am also doubtful about the wisdom of antagonising many in the north of England for such small and uncertain savings. I believe that this doubt is shared by most of our colleagues in the House." In plain English, Tory MPs feared the impact closure might have on their seats. No further argument was necessary. Mrs Thatcher consented without delay and on 11 April, Portillo wrote to the protesters' London lawyer Edward Album, with the good news that consent would be announced that day. The line was saved. His TV career was on track.

The campaign to save the S&C was without parallel in British railway history. In the end, it was a triumph of common sense, legal acumen and sheer bloody Yorkshire grit over Dr Beeching's flawed concept of a modern railway.

Given its rugged terrain and appalling weather conditions, it is not surprising that the S&C is no stranger to accidents and blockages. In all, around 90 have died and in excess of 100 injured. The worst, at Ais Gill in 1913, claimed 16 lives with more than twice as many injured. That followed the Hawes Junction crash of 1910 in which 12 died, with 17 injured. Seven were killed at Little Salkeld in 1918, two at Little Culgaith in 1930, another one at Little Salkeld in 1933. Thirty-four were injured in a derailment at Blea Moor in 1952, and five died, with seven injured in the 1960 Settle crash. One died in the 1999 Ais Gill accident. It will be noticed that a majority of these incidents took place on the remote parts of the route.

It may seem absurd in a time of climate change, but the S&C remains vulnerable to sudden, heavy snowfalls, and sustained periods of high rainfall have caused landslips that closed the line for weeks. It is a miracle of railway engineering and a tribute to the men and women who operate it, not to mention (but I will) the sterling work of the Friends of the Settle Carlisle Line, that "the line that refused to die" continues to defy the bean-counters and their dismal world view.

CHAPTER 15

STATIONS

BON VOYAGE! RAILWAY stations are not just a place to catch a train. They are romance, where you say goodbye to family and friends, or wait apprehensively for someone you love. Emotional, fascinating – it doesn't take Celia Johnson and her *Brief Encounter* to spell out the message: here is the thrill and passion of travel.

On an architectural level, our great British stations have been described as secular cathedrals of the steam age, even if they have been adapted for the modern age, like St Pancras, terminus for the Eurostar. They retain the Victorian – Gothic – grandeur of a bygone age, but are still alive and bustling, whether it's York, a vast, curving train shed still largely untouched, or the solemn classical style of Huddersfield.

Britain still has 2,580 stations and some fanatics like to visit then all. Rail worker Dave Jones, 34, of Three Bridges, West Sussex, worked out a Spring 2025 itinerary to "do" them all in six weeks – at the rate of 62 a day. "Some people call me brave, some call me bonkers, and some people call me things I can't repeat," he said, embarking on his railway marathon to raise money for hospital charities. Colleagues of the Govia Train Railway control room services manager followed his progress on a six-foot-high map, but he didn't quite make it. After 441 hours living and sleeping on trains, covering 9,846 miles, Dave called it a day on Inverness station with a haul of 2,542 stations. His favourite of the lot: St Ives, praised for its "stunning views".

We all have our favourite stations, sometimes redolent of childhood journeys to family and the seaside. My number one would be Normanton, at the bottom of our street in childhood days. If it was still there, that is. All that remains is the stub of a platform that used to be a quarter of a mile long, with a bus shelter and a ticket machine. As I write elsewhere, this was the premier junction for services from London to Edinburgh in early Victorian times. The Queen herself changed trains here, as did foreign statesmen.

All gone, and it would be a step too far down Nostalgia Street to nominate a station that hardly exists any more as my favourite – and judge how it compares with the verdicts of Simon Jenkins, doyen of railway architectural historians, whose 100 Best is the bible of the station story. Jenkins sees the station as a social phenomenon, "a public stage of human contact, ever more prominent in what is called the age of hypermobility. It is once again at the heart of British life." I say, steady on, old boy!

The history of the railway – never "train", that is an unwelcome Americanism – station offers yet another

Railway beginnings: Opening of the Stockton & Darlington Railway. 1825

Scottish origins: Scotch Express - running from London Kings Cross to Edinburgh since 1862

A new age: George Stephenson's locomotive, the Rocket

LOCOMOTIVE "ROCKET." 1829.

Train pioneers: *(Clockwise)* Sir Herbert Nigel Gresley, George Stephenson, Robert Stephenson, Richard Trevithick, Isambard Kingdom Brunel

Famous trains: *(Clockwise)* No 57 Lion. Built in 1838, Lion had a top speed of 40mph; Mallard Steam Engine Train that broke the world record for steam; British Railways 'Britannia Class' No 70000 Britannia passes St Philips in Bristol; Kestrel the most powerful single engined diesel locomotive in the world in 1970

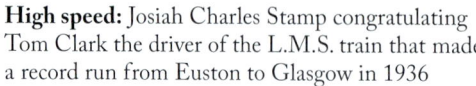

High speed: Josiah Charles Stamp congratulating Tom Clark the driver of the L.M.S. train that made a record run from Euston to Glasgow in 1936

New heights: An Air Express, the Imperial Airways latest machine, flying over another express - the Flying Scotsman at Welwyn (Herts), 1934

Rising stars: Dinah Sheridan, Gary Warren, Sally Thomsett and Jenny Agutter on the set of *The Railway Children*, 1970

Brief encounter: Richard Burton and Sophia Loren pictured at Brockenhurst Railway station in 1974

The long and winding track: George Harrison pours himself a cup of tea in his carriage pulling into Bangor Station in North Wales in 1967

Loco-motive: Star rail enthusiast and record producer Pete Waterman

Last of the summer wine: Actors Brian Wild, Peter Sallis and Bill Owen out and about at Oakworth station in Yorkshire in 1979

Festive spotting: John Crookes of Cathay, Cardiff spending his Christmas Eve train spotting at Cardiff General Station in 1959

Starting young: Trainspotters in Wales noting down the 1922 Locomotive engine *(above and right)*

To the front: Departure of The Cambridgeshire Regiment from Cambridge Station, in August 1914. Soldiers in the train and their loved ones on the platforms say goodbye

Packed carriage: on an army train taking troops back to the front after being on leave during the Second World War November 1942

Homeward bound: Returning evacuee children wave goodbye to Derby as their train moves out, bound for London and their homes, 1945

Escape to the country: Children evacuees from Birmingham arrive at Ripley, Derbyshire during World War Two

Track stars: Robert Donat pictured on the set of Alfred Hitchcock's film *The 39 Steps* in 1935. The locomotive used is a Gresley Pacific 2595

Tragedy: Some of the survivors of the Quintinshill Rail Crash near Gretna May 1915. A total of 227 people, mostly soldiers, were killed in the accident when carriages caught fire

A welcome home: Railway Guard Mrs Connelly gives a big kiss to British troops arriving home at Glasgow, repatriated under the prisoner exchange scheme. 25th October 1943

DON'T LET THE SHUTTERS COME DOWN £XCESSIVE

The future of rail: Protester outside the Department for Transport calling for plans for closure of ticket offices at rail stations to be abandoned

We can do it!: Railway inspector and women cleaners working on an engine in the floodlit pit of a train shed in 1944

Career change: Mrs. Mary Milbrand of Doncaster, who in normal times made silk lingerie, takes up signal duties on the LNER railway in 1941

Closing time: Ivor Roberts and Paul Shane from *Oh Doctor Beeching!* BBC sitcom being filmed at Arley Railway Station

A lifetime of trains: Here I am at Ingrow station, on the Keighley and Worth Valley heritage railway, standing next to LMS Ivatt 2-6-2 tank engine, built at Crewe in 1949

A trainspotter abroad: Me standing on footplate of a locomotive in Portugal, some time in the 1980s

Safari gang: Myself (in the funnel of a steam engine) on a trainspotters safari with friends in Dunfermline, Scotland, 1959. A week-long, round-Scotland trip

competition for the title "first". Purists record the first as The Mount in Swansea in 1807, but that was for the Oystermouth Railway, whose carriages were drawn by horses to the tourist delights of The Mumbles. The first used by trains drawn by engines was built just north of Darlington at Heighington and Aycliffe. It opened for passengers and freight in 1827, less than two years after the world's first public railway had begun operations. First and second class passengers entered by different doors. Toffs could load their carriages onto what would now be called low-loaders. This unique facility was in use until the 1970s on the line from Darlington to Bishop Auckland, and survives – just – today, after being declared a Grade II listed building. It fell into dereliction, but was restored as a pub. That closed in 2017 and there is a campaign to restore the station during the bicentenary year.

The world's first railway terminus was at Crown Street, Liverpool, built in 1830 for the line to Manchester, while at the other end Liverpool Road station was built soon after to serve Mancunians. The former was demolished in 1836, when passengers were transferred to Lime Street, and the latter, which looks like a row of Georgian houses, remains in use as part of Manchester's Museum of Science and Industry. I have been on a replica 1840s train ride here, but Visit Manchester says this part of the museum is closed for a multi-million restoration programme, hardly an advertisement for the railway in this of all years.

Early stations were rough and ready affairs, more suited to goods than people. Passengers could even buy tickets in pubs. But Victorian pride very quickly made itself felt on the railway. Euston lays claim to be the first in a long line of great 19th century showpieces of railway architecture, which ended with the rather more modest Marylebone a few stops away on the Underground. The London and Birmingham's terminus in the capital, opened in 1837, was entered by a giant arch with 40-foot high Doric columns, the Propylaeum, modelled on a Greek temple. The neo-classical theme continued in an impressive departure hall, described as one of the finest public spaces in London.

The entire complex was bulldozed in the 1960s, and replaced by a modernist glass structure denounced by the poet John Betjeman as "this disastrous and inhuman structure". Its ultimate fate in the great scheme of things with HS2 is unknown. Despite a public outcry, Transport Minister Ernest Marples – the politician who gave us Dr Beeching – gave permission for the Doric Arch to be demolished. Its massive ornamental iron gates survive in the National Railway Museum.

Euston, where I was an unhappy commuter from Berkhamsted for more than a decade, is a shadow of its former glory, but the railway civilisation prospers elsewhere. York, lauded by Jenkins as "the most lovable" of all stations outside the capital, is also one of my top favourites. Built in 1873 when the old station of 1841 within the city walls – breached to make way for the railway – was too small to handle the traffic, its soaring, majestic curved trainshed with glass and iron roof deserves all the superlatives heaped upon it.

And my, is it busy. It has been my good fortune to have the York Experience for the best part of 80 years. As a nipper, my mother took me there in the mid-1940s, changing trains for Scarborough. That was the destination for our family holidays for the 1950s, and it's still a thrill to take the connection from Leeds. At least once a year, I visit the National Railway Museum. But it's childhood days that matter most: standing on the eastern end of that long number 3 platform, pen and book in hand, waiting for the expresses to come round the curve from the North, clattering over the crossing for coastal services, steaming by for the waiting crowds.

York is living history. The old two-storey signal box situated right on the station, above Platform 3, with a glass frontage overlooking the pedestrian bridge across the tracks, is now a coffee shop. The station's impressive clock with three faces (one more than an MP, I've heard said) sits alongside, right above the traveller's head. Just down the platform is the station pub, the York Tap, originally built in 1872 as a tea room, still retaining its original art nouveau stained glass ceiling and windows and a stunning terrazzo floor. The beer ain't bad, either.

Next up on my list of favourites is Shrewsbury, not an obvious choice, scarcely the crossroads of the nation. But it is the gateway to the Welsh Marches, the delights of Wenlock Edge and Elgar's England, the iron road to Aberystwyth (if you like that kind of thing) and barely a stone's throw from Bridgnorth, home of the Severn Valley Railway, arguably the finest heritage line of the lot. I like it because it takes me to where I want to go, if only via the bus station: Clun, and its valley to the Craven Arms (still a station, but no engine shed), in the blessed country of poet AE Housman. The lyricist who gave us the unforgettable phrase "those blue remembered hills". Shropshire is the second-best county, after my own. I could live there and be one of Housman's Shropshire Lads.

The station is unusual, built like a castle alongside a real castle just up the hill. Simon Jenkins is quite cruel, sneering at its Tudor windows, chimneys, battlements and an entrance tower, with an extra wing that makes it look like "a miniature St Pancras of the Marches". I wonder how often he goes there. Not as often as me, I bet. A station is not simply a church without a steeple, it is the home away from home that takes you to new places and people. And the whole point of Shrewsbury is that it magics you to where your imagination wants to take you.

Enchantment of a different sort awaits me at Didcot station in Oxfordshire, on the Great Western Line from Paddington to Bristol and South Wales, 50-some miles from the capital. It is not a significant centre of population, more of a commuter railhead, as signalled in its new name of Didcot Parkway. Opened in 1844 and, rebuilt in 1985 with a 600-space car park (where I hear it is possible to fall into altercation with local Tory politicians), it owes its existence to the snooty burghers of Abingdon, which was the preferred route for Brunel's broad-gauge line to Oxford. They objected so it came here instead, spurring the growth of a small village into a railway town. The station has five platforms, an agreeable booking area, with exceptionally helpful clerks, a newsagents and sundry vending machines. Express trains rattle through at fearsome speed, while the stopping trains from Oxford take what seems an age to get to the capital.

But that is not what we are here for. Didcot station is the access point, via an underground passage, to the Great Western Society's base at the repurposed steam motive power depot a stone's throw away.

While waiting for your train, you can glimpse movements in the yard, and savour the proximity of a bygone age. Better still, visit the centre where you can walk round the shed as though you were a trainspotter in the 1950s. There is also a useful museum, where you can learn the history of "God's Wonderful Railway". If those are my top three stations, what are the bottom three?

Number one has to be subterranean Birmingham New Street. This catacomb was first opened – in the open – by the London and North Western in 1854 in a marsh area known as The Froggery. It was interred by British Rail in 1967 as part of the West Coast electrification programme, at a cost of £4.5million (20 times that today). New Street's narrow, congested platforms, often polluted by diesel fumes, are officially designated an underground station by the fire service. It counted 35 million entries and exits in the year to March 2024 with five million changing trains. BR sold what are quaintly called the station's "air rights" so it now sits below a major shopping centre, Grand Central – presumably in ironic homage to New York's historic station. It is the final surrender of the railway to retail.

Derby is the home of the Midland Railway, but you wouldn't know its history from its present. The station is a cheerless, windy site, opposite the restored former locomotive works that look an altogether more attractive place to wait for your late connection. It is sited in an urban desert, a bus ride to the city centre. The classic Midland, where I have stayed, was closed last time I was there. Derby is no friend to the traveller. Nor is my third choice, if choice be the right word. Brighton, on the South Coast, was where I lived for the best part of two years as a reporter on the *Evening Argus*. It was exciting then, the terminus of the Pullman Brighton Belle, the entry to a vibrant seaside town. Today, it is a scruffy, unwelcoming sausage machine for commuters, and it opens on to a seedy road down to the seafront that you wouldn't ask your mother to attempt at night.

Network Rail and the train operating companies share responsibility for our stations. The taxpayer funds their operation to the tune of billions, but you wouldn't know that from using some of them.

Renationalisation opens the way to a complete revamp of the system. A £45billion investment plan for the next five years was announced in April 2024. Will it run on time? Past punctuality performance is not encouraging.

CHAPTER 16

THE ROYAL TRAIN

THE PEOPLE'S EARLY fears about the risks of railway travel were shared by their monarch, Queen Victoria. Her Majesty looked on the snorting iron horse with regal apprehension, but curiosity soon got the better of her misgivings.

It was her sickly aunt Adelaide, the Dowager Queen widow of William IV, who became the first royal to travel by train, on 22 July 1840, from Nottingham to York. She obviously enjoyed the experience, because she had a special carriage built for her by the London and Birmingham Railway two years later.

Victoria's consort Prince Albert took to the train like an aficionado, but not until 13 June 1842, by which time railway travel was well-established throughout much of her kingdom, did Queen Victoria deign to give it a try. She was taken from Slough on the GWR's London Paddington to Windsor line, hauled by the locomotive Phlegethon, with none other than Sir Daniel Gooch Bt, the company's chief locomotive engineer, and engineering genius Isambard

Kingdom Brunel on the footplate. Her Majesty rode in relative comfort in a special four-wheel coach built for her in 1840.

The speed made her nervous, causing husband Albert on occasion to caution the GWR, "Not quite so fast, Mr Conductor." And for some time, the Queen required her train to observe a speed limit of 40mph, a royal command that played havoc with the timetables. But as the years went by, she became accustomed to railway travel – particularly to her new palace of Balmoral, near Aberdeen – and in 1869 commissioned a pair of coaches costing £1,800 (around £250,000 today) from the London and North Western. This was followed five years later by a royal saloon built at Swindon for her trips on the GWR and another for the London and South Western in 1877.

The first dedicated Royal Train of six coaches was built by the Great Western in 1897, to mark the Queen's Diamond Jubilee. Not to be outdone, the London, Brighton and South Coast Railway made another one, with special portions for the Prince and Princess of Wales. By royal command, the décor, whilst luxurious by the standard of the day, was "quiet and at the same time artistic". Electric lighting and toilets were now the order of the day, but the ageing Queen continued to demand a "convenience stop" at stations when she felt the call of nature. Her last journey by train was for internment at Windsor in 1901.

Her son, Edward VII, commissioned a new royal train from the LNWR in 1902, and over the next decade the GNR and North Eastern and The Midland followed suit. Wartime compelled fresh design and in 1941 the LMS built three armour-plated saloons for George VI, which he used for visits to towns and cities hit by German bombing. His train was a state secret.

After the war, and nationalisation of the railway in 1948, BR regions continued to use the royal coaches of former companies. But public pressure, orchestrated by the media, resulted in the presentation of a completely new set of coaches to the Queen to celebrate her Silver Jubilee on the throne. This modern version is a far cry from the sumptuous luxury enjoyed by the monarchs of yesteryear. The furniture is more practical – there are no double beds – and it's a more functional instrument of government, fit for a constitutional monarchy.

But that almost didn't save it from the cutter's torch. In 2013, it was suggested that the train would soon have to be axed on grounds of economy. It might go the way of the royal yacht Britannia, taken out of commission when she reached the end of her seagoing life and now a museum exhibit in Leith harbour.

Sir Alan Reid, Keeper of the Queen's Purse, told a committee of MPs that the rolling stock had only five to ten years' service left and the prospect of replacing it would be "a major decision". The figures were "quite staggering". Happily, further tests on the Mark 3 coaches showed that the train's life could be extended for many years and with efficiency savings, running costs could be reduced.

"The royal train has more life in it than was previously thought," said a royal spokesperson. There is no end in sight to its use, and no fixed, agreed horizon in terms of when it would become inoperable or need to be decommissioned."

The current royal train is composed of nine coaches painted a rich burgundy colour known as "royal claret" and is described as functional rather than luxurious in the style of its predecessors. One new feature is a security upgrade to protect the royals in the event of a terrorist attack.

Seven of the coaches are of British Rail Mark 3 design and two were originally built for the High Speed Train prototype. They boast "his and her" lounges, at first for the Queen and Prince Philip and now doing duty for King Charles and his Queen Camilla. A dining car with a table seating 12, a kitchen car, bedrooms, bathrooms and cars for staff, including it must be presumed, royal security officers and household employees.

One 75-ft long carriage for the monarch's sole use, is rather like Balmoral on wheels. The three-foot wide bed has a tartan coverlet. Beside it is a bedside cabinet with a wireless permanently tuned to Radio 4. The room is furnished with tartan curtains, while the walls are hung with framed land-scapes by the Scottish artist Roy Penny. The King also has his own sitting room.

When not on royal duty, this unique train, owned by Network Rail is normally stabled at Wolverton works, the home of carriage building in Britain, opened by the London and Birmingham Railway Company in 1838. It is maintained there by DB Cargo, which also owns the engines charged with the duty of hauling the royals round their domain. Two Class 67 locomotives, 67005 Queen's Messenger and 67006

Royal Sovereign, both painted in the same royal claret livery as the coaches, which entered service in 2003 are designated for the job, while a third 67026 Diamond Jubilee painted silver – rather like the first of Stanier's Jubilee class built for the LMS in 1934 – also joined the exclusive royal fleet after 2012. In deference to the King's well-known environmental concerns, the DB Cargo-owned engines are run on biofuel made from waste vegetable oil.

It has not always been so. Even after its demise on the system in 1968, steam locomotives have pulled the royal train in recent years. In 2005, the Duchess of Sutherland 6225 took the then Prince of Wales over the Settle and Carlisle line and in 2018 he was behind 35028 Clan Line – built in 1948, the year of his birth – on a visit to Cardiff.

Has the royal train ever been in a crash? None is recorded, as they would surely have been. The train has an excellent safety record, as has the railway in general. But there have been incidents. In October 1881, the train carrying the Prince and Princess of Wales from Ballater to Aberdeen lost a tyre from one of the tender wheels. In June 1898, driver David Fenwick was killed taking the train from Aberdeen to Perth. An inquest found that he died after climbing on to the tender on contact with a bridge, while trying to sort out an issue with the communication cord. And there was a scare in June 2000 when a royal protection officer accidentally fired his Glock automatic pistol during an overnight stop in south Wales, but the sleeping Queen and Prince Philip didn't even notice.

Driving the royal train is regarded as an honour and while there may be republicans on the footplate, there is never any shortage of volunteers. Sometimes, it's just a matter of who's around and knows the route. One such was Mick Whelan, general secretary of the train driver's union, he told the *Aslef Journal* (to which I am indebted for this story and other details). On the day in question he was a driver working out of Stonebridge Park depot on a shunt at Willesden when the driver of the royal train went sick.

Mick takes up the story, "I was asked if I had my number one uniform in my locker. I said I didn't have a number one uniform in my wardrobe! I was wearing a T-shirt and black jeans. But as there weren't any royals on board I took the royal engine down from Willesden to Euston and then took the royal train from Euston up to Wolverton." Where, reported the *Journal*, it was inadvertently derailed. "Well," said Mick, "some of the carriages went one way, and some the other..." Not quite lese-majeste, but embarrassing for the railway authorities.

Scandal of a sort has been attached to the royal train, as it has been to the royal family. In 1980, the *Sunday Mirror* reported that Prince Charles, as he was then, and his wife-to-be Lady Diana Spencer, spent several hours in a transport of delight in the train as it stood in a siding. The story was strongly denied, but rumour later suggested that "the blonde on the train" was not his fiancée but the love of his life, and now Queen, Camilla.

Like all boys who grew up in the great days of steam, the King was a puffer nutter, as enthusiasts of the era are known to their detractors. In February 2009, he named the new-build A1 Pacific, Tornado, the first express locomotive to be constructed in Britain for more than half a century, and rode in the cab from York to Leeds, a distance of just over 20 miles. His media handlers didn't say whether he took a turn on the regulator – or picked up a shovel – but we can speculate. For amateurs on the footplate, the temptation is overpowering and who was to deny the monarch his boyhood wish to be an engine driver?

Now however, a right royal romance is over. On 30 June 2025, to the dismay of the railway world, Buckingham Palace ended almost two centuries of history by announcing the end of the Royal Train. The luxury, nine-coach set will be seen on round-Britain farewell trips over 18 months, before historic parts are put on public display.

In future, His Majesty will usually travel round his kingdom by helicopter or car, though few would be surprised if he joins a regular service train. As mentioned, he has been known to enjoy a visit to the "cab", especially if it's steam-hauled, as it was again in 2018, behind 35028 Bulleid Pacific Clan Line, built in 1948, the year of his birth.

The train had been used extensively for events during the reign of Queen Elizabeth, particularly for her golden and diamond jubilees. But it found less favour with the new monarch, who has used it only twice in the past year, at a cost of almost £78,000. A two-day visit in February from London Euston to Burton upon Trent, Staffs, for a factory visit cost £44,822, while a two-day trip from Windsor to Crewe last May to visit a car plant and a charity centre cost £33,147. Axing the train will save an estimated £1million a year in upkeep and maintenance.

Making the announcement, James Chalmers, Keeper of the Privy Purse, said, "The royal train has, of course, been a part of national life for many decades, loved and cared for by all those involved, but in moving forward we must not be bound by the past.

"Just as so many parts of the royal household's work have modernised and adapted to reflect the world of today, so, too, the time has come to bid the fondest of farewells, as we seek to be disciplined and forward-looking in our allocation of funding.

"With His Majesty's support, it has therefore been decided that the process to decommission the royal train next year, in anticipation of the expiry of the current maintenance programme in March 2027.

"Before it finally goes out of service, it is our hope that the train will make further visits to parts of the UK, while discussions will begin on finding a long-term home where some particularly historic elements might go on public display."

Carriages of previous royal trains, including from Queen Victoria's reign, have been on display at the National Railway Museum in York for many years, visited by well over half a million people a year.

But it's not the same as seeing them out in the wild, and the loss of the royal train will come as a disappointment to rail-lovers everywhere.

12 THINGS YOU DIDN'T KNOW ABOUT THE ROYAL TRAIN

1. The Dowager Queen Adelaide, widow of King Wiliam IV, was the first to travel by train on 22 July 1840, from Nottingham to Leeds.

2. The first reigning monarch to take to the railway was Queen Victoria, on a GWR special from Slough, the nearest station to Windsor Castle, to Paddington on 13 June 1842.

3. The first permanent Royal Train, of six coaches, was built for Queen Victoria's Jubilee in 1897. Prior to that, each company provided a saloon carriage.

4. King Edward VII ordered two new royal saloons: one for himself, and one for his Queen, built for the London and North Western Railway at their Wolverton works.

5. Not to be outdone, the Midland Railway constructed a royal saloon for King George V at their Derby works in 1912, numbered 1910 to mark the year of his accession to the throne.

6. Three special armour-plated coaches were built by

the London, Midland and Scottish Railway in 1941 so George VI could travel around the country in complete safety.

7. Nationalisation did not mean standardisation for the Royals. BR regions continued to maintain the royal carriages of the former regional companies.

8. The first single royal train was constructed in 1977 to celebrate the Queen's Silver Jubilee. "Her Majesty loved it," says Christian Wolmar.

9. Royal train drivers are drawn from the pool of railway professionals. They are chosen for their experience, route and traction knowledge. Unflappability and skill. They must be able to pull the train within six inches of a designated station stop.

10. Original, beautiful Victorian and Edwardian carriages and mid-twentieth carriages are exhibited at the National Railway Museum.

11. Plans to scrap the Royal Train mooted in 2013 were scrapped.

12. Has the Royal Train ever gone off the rails? Yes. Legend has it that the carriages were inadvertently derailed at Wolverton works. And the driver was Mick Whelan, current general secretary of the footplate union Aslef.

CHAPTER 17

THE RAILWAY IN FILM AND TELEVISION

THE RAILWAY AND the movies were meant for each other. Movement, speed, light and dark, the actions of men and women at work or play, and later the sounds of all these things, are a film maker's dream.

Dozens of films have been set on trains, particularly in the USA where the railway opened up that great country. But the lens was trained on the industry almost as it was born.

Long before the first moving railway film shown in public in 1896, of a train entering the station of La Ciotat in the south of France, shot by the Lumiere brothers, photographers had focussed on this remarkable new feature of the landscape. One image of Linlithgow station on the Edinburgh-Glasgow line, taken in 1845 survives and the building of new lines in later years is well-documented. Bridge construction was a favourite.

But it took many more years for a moving camera to capture the vitality of rail travel. *Conway Castle*, what would now be called a travelogue, was made in 1898, filmed with a camera attached on the front of the locomotive. And the first screen smooch was caught on film on board a train in *The Kiss In The Tunnel*, of 1899, one of the earliest to edit together several related shots.

It was not until 1903 that the first proper railway movie hit the screens. *The Great Train Robbery*, a silent Western action film directed by Edwin S Porter and lasting 12 minutes, follows a gang of outlaws who rob a steam train and then flee across the mountains before being brought to book by a posse of marshals, cowboys and native Americans. This brief thriller came not long after Butch Cassidy embarked on his crime spree, robbing a Union Pacific Railroad train in 1900, surely a case of art imitating life. The plot was also based on a stage melodrama of the same name, premiered in Chicago in 1896.

From that point, it was full steam ahead, so to speak. Still in the silent era, *The General*, made in 1926, is an action comedy based on genuine events in the American Civil War. It stars Buster Keaton as Johnnie Gray, a Southern railway engineer, who pursues Union soldiers after they hijack a Confederate train. He captures the fugitives – and rescues his sweetheart Annabelle Lee held prisoner.

The climax is a real-life train wreck in Cottage Row, Oregon, watched by more than 3,000 spectators when a local holiday was declared so people could watch the "Texas" 4-4-0 locomotive plunge into a river, where it remained for

20 years as a tourist attraction before being recovered for scrap in World War Two. Keaton, who performed many dangerous stunts in the film, declared, "I was more proud of that picture than any I ever made."

Fast forward, and across the Atlantic, perhaps the greatest railway film, the first version of *The 39 Steps*, appeared in 1935, directed by Alfred Hitchcock. The story, which has been filmed no fewer than four times over a period of more than 70 years, features heart-throb Robert Donat as Richard Hannay, a fugitive from justice for a murder he didn't commit. He flees, appropriately enough, on the Flying Scotsman and escapes his pursuers from the train on the iconic Forth Bridge – an incident that does not appear in John Buchan's 1915 pioneering spy novel, but presumably was too good a shot for Hitchcock to miss.

For people like me who care about these things, the locomotive at the head of the train can just be identified as Gresley Pacific 2595, Trigo, named after a famous racehorse, winner of both The Derby and the St Leger.

The second version, in colour, of 1959 is, if anything, even more daring than Hitchcock's masterpiece. This time, the hero Hannay makes a similar escape on the bridge, but hides under the rails, on a girder dizzyingly high above the Forth. His getaway train is hauled by a selection of three Gresley A4 Pacifics in different shots, but it is 60027 Merlin on the bridge, blowing the unique "Streak" whistle before the guard shouts "right-away".

Entire books have been written on the subject of railways on film, but I have space to mention a few favourites, which may not be everyone's taste. Anyone who likes the cinema likes *Strangers on Train*, of 1951, a taut psychological thriller which has Hitchcock going back to his love of drama on the rails. But I prefer his earlier offering, black and white *The Lady Vanishes* of 1938, which captures both the tension of impending war – we have hindsight, so we know it's coming but clearly so did the great director. His later journey in *North by Northwest* of 1959 is of more interest to plane enthusiasts.

No league table of best railway films is complete with the mother of them all: *Brief Encounter*, of 1945, David Lean's three-handkerchief saga of two people thrown together in wartime, falling in love but not quite having an affair. They nearly get into the bedroom, but good old British middle-class social conscience intervenes. They go their separate ways, into the night, like the express train under which the heroine, played by Celia Johnson, almost throws herself in an agony of emotion. Strangely, she was married to Peter Fleming, a notable railway writer. You just can't get away from the issue.

Christmas wouldn't be Christmas without a showing on the box of *Murder on the Orient Express*, the 1974 version with more stars in the cast than carriages on the train. Inexplicably, given the popularity and adaptability of Agatha Christie's whodunnit detective novels, it took Hollywood 40 years to make this classic movie. The wait was worth it. Even though we all know who did it (they all did), Sidney Lumet's version is by far the superior film to the remake of 2017, even with Kenneth Branagh as Poirot and as many other

American actors as 20th Century Fox could find at the time. It didn't even have a real train or real snow. The big engine and four carriages were built on site, trundling up and down a mile-long siding in a Surrey production lot.

These are the big-screen, big-budget Hollywood greats. My top four, however, are more modest productions. *The Titfield Thunderbolt* of 1953, directed by Charles Crichton, and starring – if that's the right word, the locomotive steals the show – Stanley Holloway as the toff toper is a splendid family adventure. Inspired, so it is said, by the restoration of the Talyllyn narrow-gauge railway and the book *Railway Adventure* by LTC Rolt, it tells the story of local people fighting to keep their branch line. It was prescient, appearing in cinemas ten years before Beeching took his infamous axe to the rural railway.

In a way, the same may be said for *The Railway Children* of 1970, the first full-length movie made on a closed line undergoing restoration by volunteers. Filmed on the Keighley and Worth Valley Railway, then in the stage of reopening, its instant success hugely stimulated interest in the nascent heritage movement. The story, taken from E Nesbit's eponymous novel of 1905 (which begins "they were not railway children to begin with", but we all are) is well known: father is wrongly accused of being a spy, family move to Yorkshire to avoid scandal, the children fall in love with the railway and "the gentleman" (a railway director) who takes up their case, father is freed and Bobby, played by Jenny Agutter, utters her heart's cry "Daddy, my daddy!" in clouds of steam on Oakworth station. We took our daughters, then six and seven, to see it in Harrow when it came out, and have never stopped watching it.

Less obviously worthy of selection, but another minor English classic is *Train of Events* of 1949, starring Jack Warner as an engine driver and partly set in a motive power depot, Willesden [1A, for the cognoscenti] in north London. This "portmanteau" film, bringing together four flashback stories linked to the crash of a Euston to Liverpool Express, got indifferent reviews but still gets a showing on TPTV channel. The cast also contains a young Peter Finch and a very young Leslie Phillips, as a fireman and not yet a bounder. The live railway sequences are a hotch-potch of engines in action, including a Southern Pacific supposedly storming through Camden, but they feed a nostalgic appetite for the real thing.

My final choice is *Closely Observed Trains*, a black and white Czech film made in 1966 at the height of the Cold War. Director Jiri Menzel won an Academy Award for Best Foreign Language Film for this picture that tells the story of Milos, a frustrated teenage boy working on the railway during the German occupation of World War Two. To win the favour of stunningly beautiful Masha, he volunteers for the resistance. There are many very funny, and mildly erotic scenes, on a provincial station (actually, Lodenice in Bohemia), but the climax of Milos's suicidal mission to bomb a German ammunition train is about the most affecting I have ever seen. I could watch this film again and again, and do.

Television can never match the big screen for what they

call cinema-verite. That has always been so. When the Lumiere brothers showed their film in 1896, frightened members of the audience reportedly ducked behind their seats in horror, fearing that the train was actually hurtling out of the screen at them. Quite apart from the familiarity we now have with the railway, though it has its moments of anxiety, the telly in the corner of the room is too miniaturised, too domestic to have the same impact. That doesn't, of course, stop TV producers from trying, with variable results.

I learn from the usual source that American telly had a "blockbuster" series titled *Hell on Wheels* a decade ago, set in a mid-19th century camp during the construction of the Union Pacific across the Great Plains and featuring a cast of labourers, mercenaries, prostitutes, Cheyenne native Americans, killers and a Chinese woman named Mei who poses as a man named Wong. And others. I haven't seen it, but don't think it would bust my block.

The TV remake of *Murder on the Orient Express,* with David Suchet as the Belgian master sleuth, was first broadcast on Christmas Day in 2010 and, like the Hollywood versions, has been a feature of Yuletide schedules ever since. It's good, because nothing with him, and Eileen Atkins, and Hugh Bonneville and David Morrissey (I omit Toby Jones, because he's in everything, so he doesn't count) can ever really be bad.

Alas, it can't compare with the full-flavour versions, a test that does not mar the enjoyment of *Love On a Branch Line,* aired on the BBC in 1994 (was it that long ago?). This charming rural idyll takes civil servant Jasper from

Whitehall to a country estate once run by Lord Flamborough, a Battle of Britain hero, who lost both legs while driving a train during the General Strike of 1926, and now lives in a carriage on a private branch line. Faced with milord's gorgeous daughters, Jasper forgets his mandarin's mission and falls in love with all three. Lord Flamborough is played by Leslie Philips, who, it may be recalled, played the fireman in *Train of Events* more than 40 years earlier. What goes around... I enjoyed this four-part comedy so much I bought the DVD and you can't say fairer than that.

Finally, we must come to the Portillo Moment. No, not the night he lost his seat as a Tory MP in the election of 1997, but his rebirth as the Great Railway Traveller of the century. He's been everywhere where the train goes, or used to go, or might go, across the globe. His presentational style is a matter of taste, rather like his politics. He does his homework and he gets some interesting side stories, but apart from frequent shots of him sitting in the carriage, or the train pulling away, it's not as much about the railway as I would like. Using Bradshaw's timetable as a prop also wears a bit thin. Give me Chris Tarrant any time. His *Extreme Railways* took us to the wildest parts of the world: Patagonia, the Andes, Siberia, the Oz Outback – and Ireland. He goes looking for train trouble and finds it in the company of railway workers, with whom he has a real rapport.

The BBC has had a long romance with the railway. Series like *Railway Roundabout, Steam Railways, Dan Snow's History of the Railways*. I could go on, in fact I think I will, the railway gets you that way. *Railways, The Making of a*

Nation, The Golden Age of British Railways (it's about the nationalisation era)... and so on, and so on, but I fear the clickety-clack of railway repetition will send you to sleep so I'll switch off there. And I haven't even mentioned Netflix, because I don't have it. But you can get many railway films for free on YouTube, or via the British Film Institute website, documentaries going right back to the steam age. And no ticket to buy!

CHAPTER 18

FOOTPLATE, WORKERS

WHEN A TEENAGER applied to become a train driver in November 1977, the name on the form was K Harrison. The manager who interviewed the applicant was surprised to find they were actually a woman, Karen Harrison, aged 17. In those days – not so long ago – the footplate was a male preserve. He suggested that she might become a secretary with British Rail, but the persistent young lady was adamant: she wanted to be a train driver.

Her persistence paid off, and she was among those who opened the way for a new generation of women in the cab. Karen became a driver's assistant, as the job was then called and drove her first train out of Old Oak Common, the former GWR motive power depot in west London, on 12 November 1979.

Karen was Britain's first female train driver. Looking back later on her railway career, she said it was "ten years of hell, ten years of heaven. It's a bit tough when you're only a teenager and you're hit by this gigantic tidal wave of hate. To a lot of the men, I was the proverbial turd in the swimming pool. Every day I walked into the mess room I'd be s***ting myself, but strutting about pretending not to be. I couldn't let them create no-go areas for me; that would've established a precedent and we couldn't have that, could we? It would've been the beginning of the end."

Karen was a driver out of Paddington and Marylebone and joined the footplatemen's union Aslef on her first day, later becoming a rep and a delegate, and first female chair of Aslef's annual policy-making "parliament". Illness forced her to leave the railway, and she died aged only 50 while training to become a lawyer.

In a tribute, Aslef leader Mick Whelan said, "Karen Harrison was a pioneer – a passionate feminist and trade union activist who blazed a path for other women to join the railway at the pointy end of the train.

"Train drivers should, we think, represent the communities we serve and that's why we need more women in the driver's cab." Today, only 8% of drivers in England, Scotland and Wales are women, but 33% of new recruits at Avanti West Coast are women, and the company has a target of 50% by 2030.

Karen's heartening story illustrates the changing face of the industry's workforce. It has often been said the railway is a family and though this is an overworked figure of speech, it had some historical validity – usually from father to son. I have heard it said that getting a job on the railway was like finding a gold pig under your pillow. In the 1950s it was still secure employment, even if the pay had lagged behind newer industries. It came with a uniform and a hat, sometimes with a badge. In Normanton, railway families lived in railway houses, with rents as low as 18 shilling [90p] a week. They may have had an outside toilet, and a bath in the scullery with a table-top on hinges over it all week except Saturday night, but they were roomy, three-bedroom homes. No central heating, of course, that was only found in schools and the council offices. The only warmth came from the kitchen fire, which doubled up as a cooker for the black-leaded iron range. In winter, the frost made icy patterns inside the bedroom windows.

Everybody had the same accommodation: porter, railway widow, clerk like my father, fireman like Alf Ripley, driver like Jack Bromby at number 1 or the railway policeman next door to us. There was no false egalitarianism, but railway work was considered a cut above coalmining, which was dirty and dangerous – though jobs like shunter and plate-layer could be just as risky to life and limb.

Railway employment still had some elements of "service" about it from the earliest days of the industry. The first trade union for most grades was called the National Society of Railway Servants, a name only changed to the National Union of Railwaymen in 1913. There was never an official closed shop, except for a time in the 1970s, but pretty well everybody was a member, of the NUR or the Associated Society of Locomotive Engineers and Firemen, which had organised footplatemen (they were all men, then) since 1880, or the Transport Salaried Staffs Association [TSSA], formerly the Railway Clerks.

Aslef was easily the most militant, a cadre of skilled men who worked their way up from cleaner to fireman and then driver, a process that took years and imbued almost a military outlook.

They were the crème de la crème. A few on the LNER became "stars" even had their names on cabside plates of their regular Pacific express engines. The NUR, being a general union, was traditionally less prone to strikes, but moderation in its ranks fluctuated, depending on resentment over pay and the policies of its leaders. My father belonged to the TSSA, despite being a lifelong Conservative, until a spat with BR prompted him to join the NUR. As I recall, the first person on his doorstep in Bradford after his death in 1972 was the branch secretary with information about funeral benefits. The unions still offered friendly society benefits.

I have to confess that my own railway service was brief and unimpressive. It fell into two stints, both organised by my father. The first was as a boy assistant in the offices of the BR goods depot in Wellington Street, Leeds. My only recollection is of reading a lot of books on the firm's time. The second was more adventurous, as a relief goods porter in 1964/65. My job, if that it can be called, was to go as directed to various goods sheds in the West Riding of Yorkshire to help out with the loading and unloading of waggons and suchlike. At different times I worked in Castleford, Huddersfield and Wakefield.

The work was strenuous, though not back-breaking, but I was only a teenage lad weighing eight stone wet through.

Lugging around bales of wool in Wakefield with a lethal-looking steel hook wasn't much fun, and would probably be against 'elf'n'safety today. I also recall unloading timber from open waggons in the sunshine in Castleford and the merciless banter of the messroom in Huddersfield. "Have you got a girlfriend, then?" they teased. As if I was going to tell them. At my interview for a place at Nottingham University, I was pointedly asked if I would work in the vacations. "Yes, probably on the railway," was my innocent reply. "That's not what I mean," growled my interrogator, a clergyman-professor. He meant academic work.

I was never cut out for work on the railway, but I have always respected and admired those who do and particularly the trade unions to which they belong. They had a hard furrow to plough, in the teeth of opposition from the operating companies who regarded "combination" as a threat to the sacred principles of Victorian free enterprise. Who did these people think they were? The Amalgamated Society of Railway Servants – also known as the National Society of Railway Servants, was formed in 1872, followed by the United Pointsmen and Signalmen's Society in 1880 and the General Railway Workers' Union in 1889. This trio amalgamated in 1913 to form the National Union of Railwaymen, covering most grades except clerks and footplatemen. At its height in 1945, it had 408,900 members, but little more than half that 20 years later. In 1990, it merged with the National Union of Seamen to form the Rail Maritime and Transport Workers Union. Its charismatic leader Mick Lynch, the bugbear of the Tory media, retired in May 2025 and the RMT is now led by Eddie Dempsey. The union went into the doldrums around the turn of the century, with membership falling to only 57,000 but under the energetic left-wing leadership of Bob Crow, it rose again to more than 80,000 by 2019, levelling off just below that today.

The NUR was hugely influential in labour movement history, being one of the founders of the Labour Party in 1906, and sponsoring many MPs, including some who went on to become Cabinet ministers in Labour and coalition governments. It has wavered between moderation – in the political language of Labour – and militancy. In the 1970s, under the general secretaryship of Sir Sidney Greene, a member of the court of the Bank of England, no less, the union was definitely to the Labour Right. His successor Sid Weighell tried to keep it that way, a posture that cost him his job as the membership demanded a more militant leadership. This they found in Jimmy Knapp, a 6ft 4ins Scots signalman with a shock of white hair who learned his politics in a Socialist Sunday School in Ayrshire. Knapp, whose Kilmarnock accent was so thick you could cement bricks with it, won the top job despite his ousted predecessor calling him "a tool of the communist and Trotskyite left". Knapp invented the strategy of "one day strikes" called at unpredictable intervals until the union's objectives were achieved. The NUR's move to the left was consolidated by the election of Bob Crow and nomination of Jeremy Corbyn to lead the Labour Party.

If it was possible to earn even greater wrath of the Right-wing media than the NUR, Aslef was the union to do it. The idea of trade unionism on the footplate began spreading in the years after the first public railway, drawn together by a common bond of skills and pride of craft. The first Engine Drivers and Firemen's Society appeared on the North Eastern Railway in 1865, but was smashed "with contemptuous ease" by the company after embarking on an ill-prepared and under-funded strike. It re-emerged, with United in its title, but with more modest ambitions as a friendly society. Five more years were to pass before founder Charles Perry wrote to Willam Ullyott in Sheffield on 9 February 1880, "We want a large sum of money to protect ourselves as enginemen and firemen from the rapacity of our employers. The larger the sum we pay into it, the sooner our position will be impregnable, and once directors know that we are in reality prepared to defend ourselves, superintendents will think twice before they turn the screw." His letter is proudly preserved in Aslef headquarters. By the end of that year, delegates from depots in Leeds, Bradford, Sheffield, Liverpool, Pontypool, Carnforth, Neath and Tondu had set up the nascent union.

The Transport Salaried Staffs Association, with 18,000 members now caters for white-collar staff across the transport sector, but owes its existence to the founding of the National Association of General Railway Clerks in Sheffield in 1897. That it was established at all was a minor miracle. Two earlier attempts had failed, in the face of the usual intimidatory attitude of the railway companies and the founding fathers only decided by a majority of one vote to give it a third go. The TSSA, formally renamed in 1951, is affiliated to the Labour

Party but has generally pursued a moderate line, except when it nominated Jeremy Corbyn as leader in 2015 and spear-headed his second leadership campaign.

In the 2020s, the union was wracked with a sex and bullying scandal at head office and after an internal inquiry by Helena Kennedy KC the general secretary Manuel Cortes resigned and an all-female leadership now runs the union, with Maryam Eslamdoust as the first woman and first person of colour to be elected general secretary.

The fact that all three unions are still going strong after savage retrenchment over decades is a lasting testimony that there is another family beyond the traditional railway lineage: the comradeship of the trade union. That isn't measured in terms of how many strikes they mount – though that is always a useful index of freedom in a society – but in what their members say about each other.

Take, if I may, the obituary in the Aslef Journal of March 2025, for George Edward Rowbotham, aged 97. He joined the LMS aged 15 in July 1943, at Stockport Edgeley MPD. He survived a German V1 flying bomb destined for the town's vital rail viaduct while walking to work on Christmas Eve 1944, and went on to become a cleaner, passed cleaner, fireman, passed fireman and eventually driver.

Eddie enjoyed the shed's reputation as a "good money" depot, "even though the freight work on black engines didn't have the glamour of the London Expresses passing the shed with the [Manchester] Longsight men on their green passenger locos. Longsight had the glory, but Edgeley had the gold."

When steam ended, he went on to modern traction, but was called back to the cab in 1972 when BR relaxed the main line steam ban. And in 1980 he starred with Michael Palin in episode four of *Great Railway Journeys* on a London-Kyle of Lochalsh TV safari. 4472 Flying Scotsman was chosen for the Manchester-Sheffield, with Eddie on the regulator. It left at 10.00, like the original Scotsman from Kings Cross and with the engine's owner Sir William McAlpine, they stormed up Miles Platting Bank for the cameras. In his final days, Eddie went from messroom to classroom, where he taught fellow footplates the rules and regulations, rated as "brilliant" by his students.

He retired in 1992, after 49 years and, wrote Mel Thorley in his obituary, "He used to say he knew he was getting old when he started to attend the funerals of his old firemen. RIP, Rowy, your stories could have filled every page of this issue of the Journal." And Longsight driver Ray Bullen added, "A lovely, generous man I knew for 50 years. Eddie, you will never be forgotten."

That's the spirit of the railway I recognise.

CHAPTER 19

NOT MANY PEOPLE KNOW THAT!

OPERATION SMASH HIT had nothing to do with pop records. It was a demonstration of a train crash with a nuclear canister at high speed, to show that it was "safe".

The exhibition took place on live TV on 17 July 1984, on a disused section, the Old Dalby test track near Melton Mowbray. Using the heaviest locomotive BR could provide, a 239-ton train was run into a flask carrying atomic waste of the kind produced by nuclear power stations such as Sellafield in Cumbria.

These great steel reinforced monsters were a regular sight on the tracks, and the state-owned Central Electricity Generating Board thought it a good idea to allay public fears by showing what might happen in real life.

The engine – class 46 diesel 46009, fitted with a remote starting switch and tested to ensure it would "hit the ton' – accelerated to 100mph on an eight-mile run-up, hitting the flask sideways-on to a low-loader for greatest effect. The

179

locomotive was almost completely destroyed, with its three carriages, but the loss of pressure inside the flask was negligible.

The CEGB produced publicity material showing the impact, with lots of flames and smoke and mangled metal, but the flask undamaged. The test cost £1.6million, including lost rolling stock and compensation to landowners – taking in £2,000 for disturbing pheasants!

Presumably, the public – apart from anti-nuclear protesters who tried to disrupt the experiment - felt better after watching this theatre of the absurd. But nobody in authority explained why they thought a terrorist could find a BR loco and drive it, before crashing it at the most dangerous angle – without being stopped by a signalman with their wits about them.

* * * *

The rails of the permanent way are 4ft 8.5ins wide because that's the way Roman Emperor Julius Caesar decided they should be 2,000 years earlier.

No, I didn't believe it either. But it is plausibly argued this is so, because the earliest tramways, dating back to at least the 16th century, set their rails the same distance apart as the wheels on the carts brought over by the Romans. Caesar standardised the wheel and axle design of his two-horse-power chariots and an engineer of the 1800s, Walton W Evans, calculated that was the imperial spacing, still in use today.

But not everywhere. Other gauges are available and they didn't always obey Roman law here. Brunel's Great Western ran on seven-foot rails until 1892. India, Spain, Argentina and Russia all have wider gauges than standard and Australia had three widths until the sixties.

And there are more varieties of narrow gauge lines, including in the UK, than you can shake a shunter's pole at. Most of the proper rail countries around the world, like Japan, do what George Stephenson decreed in 1825. Stick that up your toga, Caesar.

* * * *

The army once had its own railway, at Longmoor in Hampshire, built by the Royal Engineers in 1903. For more than 60 years, it trained soldiers in the skill of constructing and running trains for war.

Operations began on an 18-inch tramway, but the line was quickly relaid as a standard gauge track. At its peak, the route from Bordon, connected to the LSWR, to the southern terminus of Liss, on the Portsmouth line, was 70 miles long.

The configuration changed, because as a training railway track was operationally torn up and reconstructed. LMR rolling stock was as varied as the troops, giving them a variety to work on.

Passenger trains sometimes served villages such as Oakhanger, Woolmer and Weaversdown, as well as Longmoor Downs, the military camp and original terminus. The line closed in 1969 when the MoD deemed it surplus to Britain's military requirements. Enthusiasts' attempts to take over the LMR were thwarted by local residents.

The railway is remembered by a huge, blue Austerity

2-10-0 LMR 600 Gordon, named after the hero of Khartoum, currently on display at the Engine House, Highley, on the Severn Valley Railway.

* * * *

The humble signal box, with its white-painted wooden sides, tall windows and large identifying sign, was once a common sight on the railway. My local station had several.

At nationalisation in 1948, British Rail inherited about 10,000 signal boxes across the system, but technological advance has transformed railway communication and they are rapidly being phased out. By 2019, only 86 were in use.

Of these, the "Big Daddy" in Shrewsbury, a 120-year-old, Grade II-listed building. The three-storey brick and wood box stands at Severn Bridge Junction, just outside the station. Signallers still operate 89 of the 180 levers dating from 1903 to give the "right away" to nearly 300 trains a day.

* * * *

There have only been three triangular stations on the rail network. Earlestown, on the line from Manchester to Liverpool, opened in 1830, and was originally called Viaduct. The waiting room on the Liverpool side is reputedly the oldest railway building still in passenger service, though invariably some will dispute this claim.

Shipley on the line from Leeds to Skipton and Bradford (Forster Square) designed by Midland architect Charles Trubshaw and opened in the late 1890's, once only had two platforms. For 100 years, Skipton trains had to reverse in and out, but a new platform built in 1979 obviated this tiresome exercise.

The third, Ambergate in Derbyshire, now only has one platform, for the trains to Matlock. Just to confuse matters, there was once a station at the village of Triangle on the Ripponden branch in West Yorkshire. Built in 1885, it closed in 1929.

* * * *

Trains that stayed in a siding and never went anywhere were once highly popular with the public. They were "camping coaches": carriages converted into holiday homes and sited at beauty spots, usually on the coast.

They were quite basic, sliding-door stock, with compartments made into bedrooms and a small kitchen with table and chairs. Moored alongside an active line, they were still on wheels and gave a new meaning to "the railway experience" – much better than modern-day "glamping".

First called "caravan coaches" and introduced in the thirties and put on hold in World War Two, they made a big comeback after nationalisation when the Western Region sited a number in the old goods yard at Dawlish Warren on the Devon coast. The London Midland put theirs at Blackpool and Abergele, north Wales.

At one time there were 230 such coaches at 61 sites around the country, but with changing holiday habits and higher expectations of accommodation, their appeal diminished and Dr Beeching's reshaping of the railway sounded the death knell.

The last camping coach left BR metals in 1971, but nostalgia-hunters can still enjoy the experience at Blue Anchor station on the West Somerset heritage railway, "updated to modern standards, yards from the beach, steam trains running by".

Aye, there's the rub. I remember a family holiday in the late fifties in a camping coach at Penally, near Tenby on the Welsh coast line of Carmarthen Bay to Pembroke. A perfect trainspotter's dream!

* * * *

The English aristocracy, owning much of Scotland as well as their own country, was frequently hostile to the advent of the railway, arguing that it spoilt their view and panicked their livestock. But not all the assembled dukes, viscounts, marquesses and plain old baronets in the House of Lords opposed the coming of the iron road.

The 3rd Duke of Sutherland, a noble family in part responsible for the ignoble Highland Clearances of crofters so sheep could profitably graze in the glens, built his own railway in the far north of Scotland.

Constructed in 1871, under his own Act of Parliament passed the year before, it ran for 17 miles along the coast from Golspie to Helmsdale, naturally with a private station at Dunrobin, to serve his family seat.

The line was absorbed into the Highland Railway in 1884 and remains an operational part of the Far North Line to this day – as does the Duke's little station, renamed Dunrobin Castle.

* * * *

Riccarton Junction was a Scottish station without a road. It was built in the Border wilderness of Roxburghshire by the North British Railway in 1862 to meet the needs of trains on the Waverley route from Carlisle to Edinburgh, where it meets the Kielder to Hexham branch.

A tiny settlement by the station was home to about 100 people, with a railway worker in each household. It had a school, an engineer's depot and a small engine shed.

The branch line was closed in 1956, and a forest road finally reached the hamlet in 1963, but the main Waverley Line was closed in 1969, amid much local anger. It was successfully reopened south from the capital city in 2015, as the Borders Railway, but far short of Riccarton.

Enthusiasts are campaigning to get a small section of restored track south from equally-lonely Whitrope, where the line reaches a summit at 1,006ft, to Riccarton. The 1,000-member Campaign for Borders Rail seeks to reinstate the entire 98-mile line via Hawick to Carlisle from its present terminus at Tweedbank, near Galashiels, 35 miles from Edinburgh.

Another Scottish station, Corrour on the West Highland Line to Fort William, the country's most remote and highest at 1,340ft, is accessible only by cycle, train or a 20-mile walk. The station house has rooms, sold out for the whole of this year. A promotional video shows what looks suspiciously like a tarmac road, but it's private.

* * * *

During the making of the 1949 Ealing Studios film, *Train of Events,* actor Jack Warner of Dixon of Dock Green, played a steam engine driver.

He agreed to the role on condition that he was allowed to drive the express out of Euston. Under tuition from a proper footplateman, he took the regulator and got the locomotive moving – not very far, and not very fast, but enough to satisfy honour.

Unfortunately, on another shoot at the motive power depot he slipped on a patch of oil and fell into the pit of an engine turntable, injuring his back and leg. The pain, and a slight limp, marked him for the rest of his days, though he lived to 85.

He was well into his fifties at the time of the filming, not particularly old for a top-link train driver. They worked into their mid-sixties, but the shed was always a dangerous place and accidents were frequent.

* * * *

All British Rail steam locomotives had a small, oval metal plate with a number and letter on the front, at the base of their smokebox. It was a code to identify the engine's home depot.

As there were hundreds of different sheds all over the country, they all had to be different, ranging from 1A - Willesden in north London, to the likes of 89C, Machynlleth in North Wales.

Before 1948, some private railway companies like the GWR and the LNWR preferred a plain, uncluttered front end, presumably on aesthetic grounds.

Such things mattered in those days, but nationalisation meant standardisation and all must be categorised. Today, these simple oval pieces of metal can fetch hundreds of pounds at "railwayana" auctions from nostalgic puffer nutters with money to spare.

Especially sought after are those from private company days, because the numbers changed after various reorganisations. For example, my home town Normanton was 20D under the LMS, but 55E when absorbed into BR's North Eastern Region. I've seen a 55E plate on the wall of the Head of Steam pub on Huddersfield station, but nary a sight of the old one since I was a lad.

By the way, the hardest shed to sneak round was 30A, Stratford in east London, the country's largest, followed by 5A, Crewe North, which was plagued by spotters. In third place, 36A, Doncaster, site of the ex-works LNER engines.

CHAPTER 20

THE FLYING SCOTSMAN – A LIVING LEGEND

THE ROMANTIC, ENIGMATIC story of the world's most famous steam engine, Flying Scotsman, has been evolving for more than 160 years. To begin with, it didn't fly, it wasn't Scottish and it wasn't even a locomotive. So, how did it earn stardom?

The narrative begins on 1 June 1862, at ten o'clock in the morning, when two trains from Edinburgh Waverley and London Kings Cross started simultaneously for the two capitals. They were not titled, but in the timetables of the three Victorian railway companies through whose territory they operated, they were known as the "Special Scotch Express". This was nothing to do with whisky, but probably a manifestation of English literary ignorance.

Over the years that followed, departure times and even routes on the East Coast Main Line changed, but the service endured and still does. Happily, it no longer takes ten and a half hours like the original run. The modern version does the job in less than four and a half hours.

More than anything, the fairy tale of the engine that refused to die begins in the fertile brain of HN Gresley – later Sir Nigel – chief mechanical engineer of the Great Northern Railway, working in "The Plant" as the company's manufacturing centre in Doncaster was known.

Ten years into the job, he decided that the Ivatt Atlantics he inherited were under-powered and could not cope with increasingly heavy trains. His solution was a revolutionary Pacific, 4-6-2 wheel arrangement, capable of pulling loads of 600 tons. Named Great Northern and numbered 1470, this monster emerged from the erecting shops in 1922. In the words of veteran railway writer Cecil J Allen, who saw the engine on exhibition at King's Cross, "To say that Great Northern provided a sensation in the locomotive world would be to put it mildly."

This was not to be the engine of history, however. The GNR was subsumed into the LNER at the 1923 Grouping, with Gresley still in charge. First results of the new design were so successful that he ordered mass production to begin. A total of 50 were built over the next two years, with the first of the new batch, 1472, rolling out of The Plant in January 1923.

There was nothing to distinguish her from her sisters, but she was chosen to represent the company at the British Empire Exhibition in 1924. Millions came to see the latest products of British industry and technology. By now renumbered 4472 and named Flying Scotsman, she was decorated with brass rims to the coupled wheel splashers, burnished tyres and the new LNER coat of arms on the cab side. A star was born.

And she couldn't stay out of the limelight. In 1928, competition intensified between the LNER and the LMS as to whose train could get from London to Scotland the quicker, the Flying Scotsman or the Royal Scot. Journey times were slashed, but Gresley devised a secret weapon to defeat his rivals. His Doncaster engineers produced a tender for the locomotive with access for the footplate crew to the leading coach, obviating the need to stop for a change of driver and fireman. The LNER's crack express could now do the entire 392.7 mile run non-stop!

Naturally, Flying Scotsman was chosen for the inaugural run on 1 May 1928. With driver A Pibworth at the regulator, she performed the task with ease, arriving in Edinburgh 12 minutes early at 18.03, with two tons of coal still in the tender. Unbelievably, she was to repeat the feat 40 years later, as we shall see.

Meanwhile, behaving as if records were made to be broken, Flying Scotsman made the first recorded speed of 100mph on the rails. On 30 November 1934, with legendary driver Bill Sparshatt and fireman Webster on the footplate, she took a 207-ton four-coach train that included a performance recording car, and "topped the ton" flying down from Stoke Summit on the East Coast main line. And this was a machine that had been in service for 11 years, covering 653,000 train miles, more than 40,000 since her last visit to the heavy repair shop at Doncaster: proof, if need be, of Gresley's genius and the skill of the workers at The Plant.

Gresley went on to design the even more successful, in speed terms, A4 Pacifics, one of which, Mallard set the world speed record for steam at 126mph in 1938. Both engines are, thankfully, preserved for all to see at the National Railway Museum in York.

But let's not get ahead of the story. Flying Scotsman continued to give sterling service, throughout the war, even if she lost her glamorous apple-green paintwork. By the 1950s, when the railways had been nationalised, Gresley was long in his grave (he died in 1941) and she was renumbered 60103, her class were slowly giving way to the new A1 Pacifics introduced by his successors.

She was still to be seen on the East Coast Main Line – I recall her at the head of an express in the 1950s – but her days were numbered.

And then fortune took yet another astonishing turn. Businessman Alan Pegler emerged as the saviour to save her from the scrapheap. As a boy, he had been taken to the level crossing at Barnby Moor and Sutton, not far from his home near Retford, just south of Doncaster to watch the expresses tearing through. He had also "cabbed" the engine at the Empire Exhibition and the experience stayed with him.

He wrote later, "Looking back, I wonder whether this was what forcibly brought home to me the difference between a cold steam locomotive and a hot one: a dead one and a live one. There seemed to be no connection between the showpiece inside an exhibition hall and the beautiful piece of machinery with its huge striding wheels, flashing side rods and plumes of smoke and steam at the head of a train."

Pegler was hooked and when his enthusiasm for the railway bore fruit in appointment to the board of BR's Eastern Region, he was in the right place at the right time. He kept an eye on Scotsman, and as she was on the point of withdrawal from active service he stepped in, buying her direct from BR in January 1963. He took her on a final run from Kings Cross to Doncaster and then into The Plant for an overhaul and some cosmetic changes. Off came the German-style smoke deflectors and a single chimney replaced the double version fitted by BR.

Under an agreement with BR bosses, Scotsman in private ownership was allowed to run special excursions on the national system, visiting cities like Exeter and Southampton that would never have happened in LNER days. Coaling and watering facilities, fast disappearing as steam traction drew to a close, were provided by strategically-stationed lorries. Pegler also commissioned a second corridor tender in 1966, giving her a range of 300 miles.

There remained one last, great challenge. Could Scotsman recreate the London-Edinburgh non-stop run, before a ban on steam haulage on BR rails came into effect in late 1971? They had to try. 1 May 1968, the 40th anniversary of the first record-breaking dash, was chosen.

On that date, renumbered 4472, Scotsman took a ten-coach train from Kings Cross amid a blizzard of publicity, with TV cameras, reporters and photographers thronging Platform 10.

BBC2 filmed the triumphal run, in which Scotsman "behaved immaculately in her best tradition," according to Pegler's fellow enthusiast Trevor Bailey. He watched the expression of wonderment on the face of a little boy sitting on his father's shoulder and pondered how he would look back on this occasion in years to come. "It was a thrilling moment and an emotional one. I was acutely aware that here was something we should never see again," he later recalled.

His fears were well-grounded, for Scotsman was about to go on her travels abroad. The Labour government gave support for the engine to go on tour in the USA and Canada, with a nine-coach train boosting British exports. A classic Harold Wilson wheeze, it ended rather like his premiership – in a flop: not for the fans, but for the finance.

Scotsman was craned onto the deck of Cunard liner Saxonia in Liverpool on 19 September 1969 for the ten-day sailing to Boston, enduring the indignity of being fitted with a cow-catcher, buckeye coupler, bell, a headlight bracket and an extra whistle en voyage, to satisfy US railroad regulations.

Her first run in October from Boston to Hartford and then New York where she was exhibited to admiring American railfans and no doubt some puzzled commuters, over four days in Penn Station.

From there, she stormed down to Philadelphia and Baltimore for more display, and thence to the capital, Washington DC, where people queued for more than three hours to get a look at this famous Britisher.

With Alan Pegler sharing the controls with a crew of

two drivers and firemen who accompanied the expedition, Scotsman took to tracks across the deep south to Birmingham, Alabama, and Atlanta, Georgia before reaching Shreveport, Louisiana and finally to Houston, Texas where 60,000 people came for the show. She over-wintered in the Lone-Star State and resumed her travels in 1970 with a marathon 15,400-mile run to Wisconsin, Montreal and Toronto, then back down to San Francisco over the Rockies. What a gel!

But she was an expensive lady. Scotsman's train was barred from charging passengers and business complaints forced her into an obscure yard where viewing takings plummeted. Pegler fell into £132,000 debt – well over a million today – with large unpaid bills and was forced to declare himself bankrupt in August that year. His beloved locomotive in secure army storage, he escaped his creditors, working his passage home on a P&O cruise ship. Ironically, he paid off his debts by spending seven years as a cruise entertainer, giving talks about trains and travel.

Scotsman was saved by another businessman, William McAlpine, who paid off the debts, bought her for £25,000 in 1973 and shipped her home via the Panama Canal. Back in Liverpool, she steamed under her home power to the BR works at Derby for restoration. Over the next few years, she worked on a variety of heritage steam railways and became the first preserved locomotive to haul the Royal Train in 1986.

But Scotsman's days of foreign travel were not over. In October 1988, McAlpine took her to the other side of the world, for a record-breaking steam safari. She travelled all over Australia, from Sydney to Perth via Alice Springs, traversing the longest non-stop run by steam, 422 miles to Broken Hill in New South Wales, and beating her own trainload record with a 735-ton train.

Returning from Down Under, BR insisted on restoring the German-style smoke deflectors and double chimney that some, including me, thought spoiled her appearance. But that was the price of her return to main line service and after a derailment at Llangollen, Wales, McAlpine had to put her into a disused BR depot in Southall, next to the GWR main line, for repairs.

She was rescued yet again by a wealthy entrepreneur and railway enthusiast. Tony Marchington, bought Scotsman, her Pullman coaches and the depot for £1.5million and spent another million overhauling her for main line running. She was back on track – with her old number 4472. Her first run out of King's Cross to York for 30 years in July 1999 attracted more than a million spectators and starred in a Channel 4 film *A Steamy Affair*. Once again, she lived up to her reputation as an expensive mistress.

Tony Marchington's plan for a Flying Scotsman Village in Edinburgh, came to nothing, and his business venture with the locomotive failed with debts and an overdraft of nearly £2million, leaving him bankrupt in 2003.

There was nothing for it but to put the ageing lady – now nearly 80 – up for sale to clear the debts. Up with this, the public and the heritage movement would not put. Rather than allow this national treasure to fall into foreign hands, the National Railway Museum bought Scotsman, plus a load of spares and a support coach, for £2.3million, with money mostly from a £1.8million grant from the National Heritage Memorial Fund. She arrived in York for Railfest in June 2004, where I saw her again for the first time in almost 50 years. She went into the NRM's own workshops for an overhaul in 2006, estimated to cost £750,000 and take 12 months. £4.3million and almost ten years later, she returned to steam in January 2016 on the heritage East Lancashire Railway. I was there, too.

She made a royal return to Kings Cross in February and her run to York was forced to a stop because spectators crowded on to the line near St Neots. A poignant memory was celebrated in October 2018, when the ashes of Alan Pegler, who had died six years earlier, were deposited in Scotsman's firebox as she climbed Stoke Bank, to the accompaniment of a blast on her whistle.

Back in the workshops in 2022 for yet another overhaul for her centenary year, Scotsman is now certified to run on the main line until 2029, after which she will be confined to heritage railways until 2032.

But the story rolls on, and on. In 2023 Scotsman appeared on a collectable coin issued by the Royal Mint and on a set of Royal Mail stamps, the last to bear the head of Queen Elizabeth. Poet Laureate Simon Armitage has penned a poem *The Making of Flying Scotsman.*

She was also operated by an all-female crew, though not on 29 September 2023, when Scotsman collided at low speed reversing on to her train at Aviemore on the Strathspey Heritage Railway.

The world's least-shy engine has appeared frequently in film and on television. As early as 1929, she starred in a film bearing her name and in *Agatha*, 50 years later. She even made a cameo appearance in *101 Dalmatians* in 2000, leaving the old Midland terminus of St Pancras, an unheard-of heresy. On TV, she appeared in a BR advert, and, incongruously as a model in a James May programme. Two further documentaries followed in 2016, and more are certain to come in this bicentenary year. For good measure, she stars in a *Thomas & Friends* film, on a long-playing sound record and on a computer game.

All this glamour, for 98 tons of steel and steam! NRM curator Bob Gwynne puts it beautifully, "Flying Scotsman is an icon, an equal to Big Ben or the Spitfire. She was the pinnacle of engineering of her day, but 100 years on, she still has class, but combines that with somehow being a nostalgic symbol of a rather warm and friendly not-too-distant past, a simpler and happier time. It's strange but true, Scotsman has now been in preservation longer than she was in actual service."

CHAPTER 21

THE RAILWAY IN BOOKS

TO MY MIND, there is something inevitable about the attraction of the railway to writers. All those words between travellers and trainmen, all that movement, the rapid changes of scene, all those people, going about who knows what business, all those sounds, smells and mystery.

The railway is a natural backdrop for fiction, but it is equally at home in verse, non-fiction and even figures. What more inviting volume can there be than the 1955 timetable for Yugoslav trains given to me as a Christmas present by an understanding daughter? Fascinating! Services to Sarajevo, Belgrade, Nis, Bar and Dubrovnik, and probably a work of fiction anyway, in my experience of Balkan railways.

The same was sometimes said of the British Rail complete national timetable, a 2,672-page page monster printed on Bible-thin paper. A copy of almost the last edition (it was discontinued in 2007 due to "lack of demand"), for Jan-June

2005 and priced £12 (£25 today), has a place of honour in the bookcase in my downstairs "loobrary".

The first public train service was breathlessly reported in the newspapers on the day after it began operation and within the decade, the railway commanded the attention of artists in prose. They were building on a strong tradition. Fictional journey narrative dates back at least to Homer's *Odyssey* and by the 18th century, the novel of travel was an accomplished form.

Gulliver's Travels of 1719 may have been by coach and ship rather than the iron road, but they fed the appetite of the armchair traveller, as did Tristram Shandy and Joseph Andrews. Daniel Defoe's *Journal of the British Isles* was supposedly a true record, but it reads like imaginative recollection, if not in tranquillity, then in the pub or coffee house.

Connoisseurs of the genre generally begin with Dickens, the survivor of a train "smash" in 1865 as the Victorian called these incidents. But go back a little earlier and in November 1829 we find the diarist Thomas Creevey recording "a lark of a very high order". Invited by the gentry of Knowsley, Liverpool, to ride behind the Loco Motive machine, "I had the satisfaction, though I can't call it a pleasure, of taking a trip of five miles in it, which we did in just a quarter of an hour – that is 20 miles an hour."

He found the speed frightful. "It is really flying and it is impossible to divest yourself of the notion of instant death to all upon the least accident happening. It gave me a headache which has not left me yet."

The choleric scribbler declared himself glad to have seen "this miracle" and travelled in it, but having satisfied his curiosity, this first achievement would be his last.

Others were more robust. Writing in July 1837, Charles Greville found the velocity "delightful". After a trial run, he concluded, "It certainly renders all other travelling irksome and tedious by comparison. It was particularly gay at this time, because there was so much going on. There were all sorts of people going to Liverpool races, barristers to the assizes, and candidates to their several elections." There you have it: the drama, the movement, the variety. It was an intoxicating brew.

Dickens never quite got over his dislike of the railway – a mad dragon, he called it on his travels in the USA and his 1866 short story *The Signal Man* reads like a kind of exorcism. It portrays the nightmare of a signalman, who sees a ghost waving to him from the track, knowing that a crash will ensue every time the spectre appears.

But by 1871, Lewis Carroll can set a child's fantasy in a railway carriage in *Alice Through the Looking Glass* and by the end of the century, Sir Arthur Conan Doyle puts Sherlock Holmes on a train in most of his stories about the famous detective. They regularly show up on TPTV, the classic film channel.

Most famously, Agatha Christie's Belgian detective Hercule Poirot solved the *Murder on the Orient Express* in 1934, a whodunnit that has spawned at least three major films and I don't know how many imitations. For my money, however, her 4.50 from Paddington in 1950 that featured her other creation, spinster sleuth Miss Marple, takes the palm – for the dramatic shots of GWR trains in the TV version.

Bringing the genre up to date, Patricia Highsmith's novel *Strangers on a Train* of 1950, was made into a Hollywood film by Alfred Hitchcock, as we have seen. And in recent years, Andrew Martin, son of a York railway executive, has pleased readers with his series of detective novels starring railwayman Jim Stringer. His male Miss Marple is a lot cleverer than the railway police sergeant who lived next door to us in Railway Terrace.

Those are the main lines of railway literature. However, I must take on the branch lines, dear reader, by which I mean the writings of the drivers, firemen, signalmen and sundry other railway workers themselves. The story from the inside, on the footplate, real and fact-based fiction. These are the words you don't hear in the ballyhoo of showbiz.

Let's start with David L Smith, a collector of stories of drivers on the old Glasgow and South Western Railway from the 1880s down to the 1920s. His remarkable tales of a rollicking life on the footplate began appearing in *The Railway Magazine* in 1939, and were collected for publication in 1961 as *Tales of the G&SWR*. His is the most entertaining and informative account of "the old days", told as you might hear them in the depot after a hard road to Stranraer on 205 "a grand old engine and in her day a flyer, as she had every right to be, for she was one of James Stirling's famous 4-4-0's of 1873". It's railwaymen's talk, of Wee Willie, Old Bob Scott, Jimmy Williamson whom they called The Drover, and a terrible big wild man they called

The Mool, and their hair-raising exploits at the regulator of open-cabbed express locomotives, night and day, winter and summer. I can almost smell the smoke of their pipes in the bothy.

Railwaymen were thoughtful people in those days. I have only a few of the books they wrote, but I count among them Harold Gasson's *Firing Days, Reminiscences of a Great Western Fireman,* published in 1973, his attempt to recapture "the happy days of steam" served with a most Honourable Company of Gentlemen – the enginemen and shed staff of Didcot Locomotive Department, working on "God's Wonderful Railway." Didcot, on the Paddington to Bristol main line, was not an express depot, worked many branch lines in the region and saw sterling service in World War Two.

Harold's beautifully illustrated autobiography is a delightful narrative of growing up in a railway family, doing his time as a cleaner, fireman and a driver – taught by his father, also HH Gasson. His days in the cab during wartime make fascinating reading for steam aficionados. I was also amused by the copy of a letter to fellow-fireman Jack Peedell of Old Oak Common depot from the Motive Power Superintendent, thanking him for his duty on the Royal Train on 26 March 1955. To mark the occasion of firing Her Majesty, so to speak, from Shrewsbury to Windsor he was awarded a gratuity of ten shillings [50p]. The driver would have got 20.

Another fireman – also GWR – Tony Barfield of Kidderminster depot published *When There Was Steam* in 1976, a collection of anecdotes from the 1950s footplate, all true.

Well, as true as anything is when retold after the event. But you can't fault the honesty of prose such as his reaction to being given "old 29" to work the thrice-weekly goods from Cleobury Mortimer to the Admiralty Depot at Ditton Priors because his regular engine had dropped a brick arch into the firebox, "29! We'll be lucky if she gets to Cleobury light engine never mind the goods."

And to the loco, "Well, old lady you are in for a treat today, you are going home back up the Gadget – the local name for the branch. On hearing such wonderful news 29 promptly blew off at the safety valves!" This long forgotten line, closed to passengers, in 1938, and the loco was fitted with an unusual chimney cowl for working into the naval depot. "I really had a soft spot for old 29," recalled Tony. That was how they talked of their engines, and always a she, like a ship.

Railwaymen usually wrote about their own work. Peter Kirton offered a unique insight into the world of signalling with his life story *Proceed At Caution*, published in 1998. A miner's son, he started railway life as a signal lad in March 1951, recording trains in Goose Hill signal box in Normanton, West Yorkshire (and coincidentally my home town). He worked in boxes around the area, and relates the often lonely, highly responsible life of the signalman, attending to his bells, levers and electric communication. It could also be tragic, as when his fellow signaller down the line observed an express running down a contractor's man right outside his box. In those days, there was no counselling, as there is now.

Kirton moved up the grades, becoming a travelling ticket inspector and a senior conductor, occasionally handling Royal journeys on the scheduled trains. He didn't forget having to fork out for refreshments for Princess Margaret's party between Peterborough and Kings Cross. "I didn't spoil the Lady in Waiting's day by asking her for my £7.83," he recalled, adding with classic Tyke humour, "although it did take me three weeks to get it back on expenses."

It was unusual for railwaymen to stray into fiction – their work was far-fetched enough – but Raymond Flint did so with his novel *Men Of Steam*, set in Scarborough – fictional Castleborough during World War Two and the post-war years. It tells the story of Joe Wade, a driver on the LNER, just as Flint was himself for 22 years. Urged by friends and family, "don't just write about steam locomotives, write about them and the people that make up the railway experience." He does that, in a novel of working-class Yorkshire folk, as well as an acutely-observed account of life on the footplate. It's an absorbing read and all the more remarkable having been written by a man suffering Parkinson's with help from the charity of the disease.

Railwaymania wasn't confined to workers in the industry. For reasons never quite explained, the railway was also a source of inspiration to "gentlemen of the cloth" (there were no ladies in those days). I pass over the children's classic, the Reverend Awdry's *Thomas the Tank Engine,* that has brought joy to generations of kiddies, but I must mention Roger Loyd's *Farewell To Steam,* of 1956, a lament for the locomotive like no other. "It is common knowledge," he writes "that many hundreds of clergymen are enthralled by everything that has to do with railways." But this second calling is difficult, for the parson must pursue what engines run through his parish only in his spare time "since it was not for this that he was ordained".

Canon Lloyd managed, nonetheless, not only to live in five different parishes (one the cathedral city of Winchester) of railway interest, but to write three highly-regarded books about his earthly passion. Born in Eccles, Lancs, and ordained in the Anglican church in 1924, he was influenced by Christian socialism and the relations between workers and religion. He died in 1966 but his work still appears in anthologies.

Finally, we must consider the politicians. They, too, are afflicted with the railway writing bug. Surprisingly, not the rail-union-sponsored MPs, though there have been dozens, if not hundreds, since the Labour Party was created in 1906, but latter-day Conservatives.

Step forward, Robert Adley and Sir Gerald Nabarro, honourable members for Christchurch and South Worcestershire respectively, and Puffer nutter (South) simultaneously. Adley had an obsessive love of trains from the age of three, but unlike most men he didn't grow out of it. He published no fewer than ten books, mostly with "steam" in the title and chaired the Commons Transport Committee of MPs. He slated his own government's plan to privatise British Rail as "a poll tax on wheels", a prediction that was proved correct but not until after he died aged 58 in 1993. His books were total-trainspotter, replete with emotional detail. I once owned three, but donated them to the Keighley and Worth Valley second-hand shop at their Ingrow Museum, which is worth a visit just for the books.

Sir Gerald published a book about the Severn Valley Railway in its infancy as a heritage line in 1971, narrating its history and confessing that it was practically in his blood. Reared close to the GWR main line a few miles west of Paddington, he recalled his first ride behind a Deans goods in 1926 and his adoption meeting in Kidderminster presided over by Sir Francis Winnington, great grandson of the line's founder. A second, *Steam Nostalgia,* came out the following year.

"Nab", sporting a Jimmy Edwards-style handlebar moustache and driving (occasionally) a car with number plates NAB1, was much-favoured by newspaper cartoonists. He fought the closure of the line and became a director of the heritage SVR. His book may be a homage to its survival, but the railway volunteers threatened to strike after discovering he planned to sell the Bridgnorth station site for development. He resigned in 1973, there were no more railway books and he died that year aged 60. "Nab" wanted to become the Hudson of the heritage movement, but he was no George.

CHAPTER 22

GREAT RAILWAY JOURNEYS

IT TOOK THE pointy-heads at the BBC a long time to work out that the railway journey is a terribly romantic story, with lots of cinematic potential. The outcome was the hit series, *Great Railway Journeys,* starting in 1980 and popular on iPlayer. Celebrities were sent on all-expenses-paid trips, usually to exotic places in parts of the world that had curious or impressive services through inhospitable terrain. Preferably with lots of ruins.

Playwright Michael Frayn went on "The Long Straight" from Sydney to Perth, TV historian Michael Wood took the iron road from Cape Town to the Victoria Falls, actor Stephen Tompkinson travelled from Singapore to Bangkok via the Malaysian capital Kuala Lumpur, a trip I sometimes took when I was the South-East Asia man for *The Times.* Alexei Sayle rode from Aleppo in Syria to Aqaba in Jordan, fortunately before civil war made that part of the Middle East a no-rail zone.

I have done my best to join this magic circle of expeditionary travellers, most recently in a 12-hour journey from Zagreb to Sarajevo, on a train through "enemy" territory of Republika Srpska where my passport was taken away for examination. The service began with a single carriage and underwent frequent changes of motive power. At one point I'll swear the train had more engines than coaches.

And I've entrained in the USA, in Egypt [Aswan-Luxor, still have the ticket], in Australia, Japan, China and virtually all of Europe with the exception of Albania, where I'm not certain there are still railways and from Athens to Kalamata in the Peloponnese of southern Greece, where there are definitely no trains.

Here I go again, this time for my annual train safari in Scotland with my longstanding pal Charlie Whelan, one-time Press Adviser to Labour Chancellor Gordon Brown. He stays (as they say hereabouts) near Aviemore, practically within sight, and certainly within sound, of the Strathspey heritage railway. As every good journey should, it starts with a swift sherbet to say hello, on this occasion in Dow's Bar, next to Glasgow Queen Street. The station has had a multi-million-pound facelift, with a soaring glass frontage that makes it look like the headquarters of a dodgy finance company. The inside is as inhospitable as ever, with few seats, a tiny, dingy waiting room and no bar. If there is a top ten of stations to avoid, then Birmingham New Street would be at the top and Queen Street not far behind.

Greetings concluded, we're off on the Iron Road to the Isles, aboard the 12.22 to Oban, a fishing and ferry port on the West coast that's become a popular tourist spot with discerning travellers. There's not a lot there, apart from a giant folly built like a coliseum on the overlooking hill, but the train journey there has few to match it. We're on a six-coach diesel, with only the front two for our destination, the rest going to Fort William and Mallaig, another great railway journey.

But we did that a few years ago and we're up for a more adventurous return to Oban, this time with a sea trip to the Western Isles, where there might just be a hidden railway delight.

Immediately out of the station, like a greyhound out of the trap, the train is on full power, charging into a tunnel up the mile and quarter-long Cowlairs Incline, a gradient so steep it was worked with a stationary engine and cables from when first opened in 1842 as part of the Edinburgh to Glasgow railway until 1908. The climb goes up at 1 in 51, then 1 in 43 and finally 1 in 41, the most spectacular exit from a main line station in the country – or would be, if most of it wasn't in a tunnel.

This is just the shape of things to come. This part of the West Highland line was not only a great challenge to build through inhospitable terrain, it also had a troubled history. First attempts to open up the area with a Glasgow to Inverness route via Glen Coe was bitterly opposed by shipping and rail rivals and failed in 1883. It was another six years before construction of a new scheme began, marred by legal and labour problems and it was not until 1894 that trains began running. But our line into Oban was originally part of the Caledonian Railway from Perth through Callander, opened in 1880. If that was confusing then, it still is. Our train divides at Crianlarich and we glide down to the old "Caley" route to the coast.

But let's not get ahead of the train. Leaving the suburbs of Glasgow, we power along the north shoreline of the Clyde coast, bowling through Bowling, where the east-west Junction canal ends. On past the Erskine road bridge we see beaches fouled with every sort of plastic and wooden debris, and flocks of eager little oyster catchers dipping and darting in the mud. Across the river lies the town of Port Glasgow.

The modern-looking vessel moored there, Charlie tells me, with grimly-humorous satisfaction, is one of three ferries ordered by the SNP government of Scotland. These ships, fantastically expensive and equally late, have yet to enter service, and the scandal of their construction may yet do for the "Nats" in next year's Holyrood election.

Nearing Dumbarton, we see a lonely obelisk dedicated to Henry Bell, the pioneer of steam navigation in Europe, says my other companion on this trip, a guidebook to the line, published in 1893 for intrepid Victorians. A little further on, also on the shore, comes the impressive Dumbarton Castle atop a 250ft-high basalt rock – the Gibraltar of the Clyde. Appropriately enough, when Scotland had its own navy, it was based here. Many are the stories of heroic sieges and assaults. Here, Mary Queen of Scots departed for the French court and William Wallace was betrayed to the English.

But let's take our nose out of the history book and look at the country scene unfolding as we pass through Helensburgh, our last station on the Clyde and sometimes called Glasgow's Brighton, "the favourite resting place of the busy city man, composed of villas for summer occupation". Given the state of the south coast resort last time I was there, this is not a title the estate agents might use today. The town, named after the wife of its founder in 1777, landowner Sir James Colquhoun, sits on the southern end of Gare Loch. Eminent Scots rather liked naming towns after their family, a habit they took to Ireland.

It is a marvellous day, very warm and not a cloud in the clear blue sky. "We don't get many like this in the summer, much less April!" observes Charlie. It's the first of the month and I had thought of doing an April Fool trick, but it's past midday now, so too late. All ScotRail stations, I note, have dual-language names in English and Gaelic.

Dumbarton Central is Dun Breatann Meadhain. A proper mouthful. But why do they do it? Gaelic may be heard where we're going (and I do hear it) but it is little spoken in the Central Belt, where most Scots live.

We're sitting on the correct side, for the mountains begin to rear up on the left of the train. Within minutes, the massive submarine base at Garelochhead comes into sight, far below the line, a military blot on the landscape. Brightly illuminated, forbidding in size, it shows little sign of activity. Perhaps it's all underwater. A solitary CND placard sits in a field by the line. This has long been a refuge for shipping. The loch, an arm of the Firth of Clyde, is six miles long and a mile wide. "So still and deep are its waters, and so free are the surrounding hills from magnetic influences, that it is the

211

favourite place for the testing of ships compasses," says my 1893 guide. Ah, so this may be why the US and UK navies chose Gare, of all Scotland's deep-water lochs, to base the nuclear deterrent.

From here, our train crosses a narrow neck of land to run alongside Loch Long. Trees, mostly ugly fir plantations, take the place of crofts. A new gravel road, seemingly to nowhere, but obviously to take out timber, scars the hillside. Where the trees have been felled, stumps shine white in the sun, like gravestones. "If it be possible, Loch Long is even more beautiful than Gareloch, and it is infinitely more impressive," says my guide. The book, that is, not Charlie, who is deep in concentration looking at the mountains which rear their brown snouts and peaks into the clear Spring air.

At Arrochar and Tarbet, we rattle down to the "bonnie banks and braes" of Loch Lomond, Scotland's largest, 14 miles long and up to five miles wide, celebrated in song and poetry. Its sheer size – the water covers a land area of 20,000 acres – intimidated the poet Wordsworth on his visit here with Coleridge in 1803. He thought "the proportion of diffused water was too great", hardly a fitting sentiment from the writer who practically invented mountain-tourism in his native Lake District. A long-distance path skirts the shore of this inland sea – which reminds me of Lake Baikal – passing below Ben Lomond, to the 3,196ft peak of which rises a stone staircase. It is the most popular "Munro" (3,000ft-plus) mountain in Scotland and Charlie is reluctant to say he's actually "bagged" it. "I nearly climbed it once, 40 years ago," is all he will admit. He has however "walked over the tops" and taken the waters at the isolated Inversnaid hotel on the far shore from our train, so honour of a sort is satisfied.

Our train hugs the western massif of this rightly-popular jewel of the Highlands all the way to Ardlui (Aird Laoigh) at the head of the loch. We stop here for a train to pass by – this is a single line, controlled by token operation – and some of the passengers step out for a cigarette. Smoking seems more prevalent in Scotland. I'm more concerned at the appearance of wild rhododendrons that are now so common by the trackside and on the hills. This invasive plant, brought here from Spain or Asia by Victorian toffs who wanted a splash of colour on their estates, plus shelter and game cover, has blighted parts of once-wild Scotland. So much so that in 2020 the National Trust for Scotland began Operation Wipeout, aimed at freeing the country's natural environment of this pretty, but unwelcome scourge. Good luck to them. Selective breeding by landowners has produced an exceptionally hardy hybrid that loves the climate here. "When you see it, always look for the big house of the bastards who brought it here," says Charlie. And he's right. It's usually there, discreetly in the background.

I realise that there are some things missing, by comparison with the Pennines. Lots of sheep, where grass can grow, but no lambs, where there are plenty in my home country. Not yet, it seems: too early. And no, or very few, stone walls rambling up the hillsides. Some stone enclosures, but no epic drystone limits up to the horizon.

That, apparently, is because there are so few landowners,

and so great are their holdings there is no need to divvy up the landscape into fields. The sheep can run virtually wild, like the deer, which actually does.

After taking two hours to do the 50 miles from Glasgow, we arrive at Crianlarich (A 'Chrion Laraich), where the West Highland intersects the old Callander and Oban. This village, which my 1893 guide says "used to be regarded as the limit of civilization in these parts" sits below the dreadful grandeur of Ben More, 3,843ft and Stobinian, 3,827ft to the east, and to the west Ben Dhu-Craig 3,209 and Ben Oss 3,376ft. My guide likens this region to Kipling's little kingdom on the road to Tibet, which was only four miles wide "but most of the miles were stood on end owing to the nature of the country". I prefer Czech humorist Karel Capek's description of Switzerland, as "a vertical country, which if smoothed out, would take up much more room".

The train divides here. So Charlie uses the opportunity to ask the conductor for some travel info. We're due to take the ferry from Oban to the Isle of Mull, and the return sailing on the following day gives us less than ten minutes to disembark and catch the train. "Does the train wait for the ferry?" he asks. "No," says Mr ScotRail. "We don't wait for the ferry, and the ferry doesn't wait for us." So much for the Integrated Transport Policy the politicians talk about. It will have to be an early start back.

No matter. We clickety-click, diddle-de-dum into Strath-fillan, no long-welded rails here, thank you very much, one of the earliest routes of Christianity into Scotland, with St Fillan from Iona. Here was a Holy Pool that cured the sick and a chapel named after the saint. We must have been blethering, or daydreaming, because I missed this important view and the site of the Battle of Dalry, a few miles further on, not to mention the old lead mines once worked by Lord Breadalbane. Blink and you lose something in this country's over-abundant history.

Leaving Strathfillan, we enter the valley of the river Lochy, at Tyndrum, a place I remember from my trainspotting days. Here, in 1959, leading a trio of Tykes on a week-long, round-Scotland trip much more ambitious than today's modest safari, I engineered a timetable switch from the Upper to Lower station (it could have been the other way round) to fit in Oban and Fort William on the same day. It worked.

We tumble down the valley to Dalmally (Dail Mhailidh), a perfect little station, with a white picket fence, flowers, and a graceful statue of a bird that Charlie says is a heron. And who am I to quarrel? He's the naturalist, protecting wildlife in the woods round his home in the Cairngorms. When, that is, he's not up to his neck in the river Spey trying to catch it, being a keen fisher of the salmon. And that's an understatement.

From here, we rattle down to the head of Loch Awe, with 13th century Kilchurn Castle, once the stronghold of the Campbells of Glenorchy for 150 years, sitting in the water on a rocky finger of land. Had I known its lineage, I would have chanced a crack about Alastair Campbell, Tony Blair's spin doctor and Charlie's great rival in the 1990s, but I didn't so the moment passed. And the sheer majesty of the mountains crowding about was enough to compel silence. No wonder they call Scotland's longest freshwater lake Loch Awe. It is awesome, though in fact, it's just Obha, in Gaelic.

Rising and falling as the gradient shifts, we come to Taynuilt (Taigh an Uillt), with ornate baskets of flowers on the station, and a boarded-up signal box. The streams and rivers now run west into the Irish Sea, and our train rattles along Loch Etive to Connel Ferry, where there must have been a ferry, for the scenic drive to Ballachulish and eventually Fort William. But the handsome cantilever bridge over the loch originally built in 1903 to carry the railway closed in 1966 and is now used for road traffic.

I think I've seen it before, though that might have been in a scene in the 1981 film *Eye of the Needle*, in which Irish rebel Donald Sutherland rides across on a stolen motorbike and throws it in the water after it runs out of petrol.

It's all glamour, round 'ere. Except when our train reaches Oban at 15.45, rolling into a nasty, featureless yellow-brick shed, plonked on the site of a glass-roofed wooden station built in 1880. This arts and crafts treasure boasted a 270ft-long trainshed, topped by a French-style clock tower visible from the sea. This imposing structure, a proper terminus for a Victorian railway, with a circulating area for hundreds of passengers, had a licensed buffet, paper shop, offices, parcels despatch and waiting rooms, and was built a few yards from the harbourside where MacBrayne's vessels departed for the Isles, as they do today. Oban was advertised as "The Charing Cross of the Highlands" by the Caledonian Railway. The new station, less attractive than many a public lavatory, was built in 1986 and the original fine art creation demolished the following year. Its loss is still mourned.

My objective was beyond this architectural banality, however, and after a night in the resort we embarked on the MV Coruisk (Coir Uisg), built in Devon in 2003, a drive-through ferry, for Craignure on the Isle of Mull. Here, according to an old restored railways guide of mine long gone to Oxfam, and a website of secret Scotland, there was a charming little narrow-gauge, steam operated railway, running from the ferry pier to the imposing Torosay castle, a country house built in Scottish baronial style back down the coast. With a 101/4 ins (260mm) gauge, one and a quarter-mile long, the Isle of Mull Railway was opened in 1983, by castle owner David Guthrie-James and local business-man and railway enthusiast Graham Ellis.

The line was a success, but its future was in doubt as soon as the castle was put up for sale in 2010 and opening days reduced. The last service ran on 1 September 2011, hauled by locomotive Victoria. The line was closed, the track lifted and the rolling stock sold, mostly to the Rudyard Lake railway in Staffordshire. The then laird of the property, Christo-pher Guthrie-James, sold Torosay, having told a Scottish newspaper he had "let them play trains for 30 years free of charge".

This unique Hebridean island line had six engines, three of them steam: Lady of the Isles, Waverley and Victoria, and 12 passenger coaches. It appeared in an episode of the CBeebies children's programme Balamory, based on the island's "capital", Tobermory. So I was too late. No trace of this charming attraction remains – except, that is, in Tobermory's seafront museum, where archivist Georgia Satchel showed me a large cardboard file of memorabilia. Faded newspaper cuttings, unused Edmondson-style tickets, a publicity brochure of this 20 minute ride through "rare scenic beauty": a treasure of lost artefacts of the first – indeed, only - railway of the Isles. "I only went on it once," Georgia tells me. "I was really upset when it closed." That's the kind of emotion the railway evokes, very often in people who know little of it.

Tobermory itself, the end point of our journey, is reached by an hour-long bus ride from Craignure along the coast road, much of it single-track. I wouldn't say that time stands still in this tiny harbour resort, but it certainly isn't in a hurry.

The road along the seafront ends in yet another jetty – these islands have them like people have fingers and toes – with yet another ferry, this time for Ardnamurchan. As a fan of the Radio 4 Shipping Forecast, despite it being shifted to a later time first thing o'the day, I fancied a half-hour trip across the Sound of Mull to the place I hear about most mornings. But Charlie is reluctant, and I drop the idea – only to learn later that the three-car ferry goes to a nowhere place called Kilchoan on the eponymous peninsula, not the mysterious Ardnamurchan Point to Cape Wrath of broadcast fame.

The trip back was uneventful, though it took me four trains and nine exhausting hours, ending in the dark with a steep hill to climb.

That's what trainomania is all about. The high point of day two was an hour sitting in the sunshine outside the Corryvreckan, Oban's Wetherspoon by the harbour. This great hangar of a pub is built on the former Railway Quay, land reclaimed from the sea for the lost, lamented railway station. Still, there are compensations. A pint of Belhaven cost £1.79, almost exactly a third of the price in Dow's Bar, Glasgow. Some journeys end where they begin.

CHAPTER 23

CRIME AND COPS

THE RAILWAY BROUGHT new freedom for men and women to travel far from their home, in family groups or singly. It also brought a hothouse of proximity between complete strangers of both sexes, and close contact with prisoners in transit. What could possibly go wrong?

By the 1850s, passengers were becoming used to sharing their carriage with a policeman in uniform escorting a man in handcuffs to a remote spot such as Dartmoor prison. But they had more to fear from the card-sharps, petty thieves and even murderers than those who were detained at Her Majesty's Pleasure. That there were so few assaults, in relation to the millions of travellers, was probably due to the large number of staff on the trains and platforms. The railway was always a labour-intensive industry and you were never very far from someone in uniform.

The first recorded murder in a carriage took place in London in June 1864, when Thomas Briggs, aged 69, chief clerk of a bank, was found dead on the track of his commute home to Hackney from Fenchurch Street, killed by a mystery assailant and robbed of his gold watch and chain and personal items including a fine beaver hat. The hat was the miscreant's undoing. A cabman recognised German migrant Franz Muller wearing it and informed the police who gave chase. But their quarry escaped by ship to the United States, from where he was eventually repatriated, tried, convicted and hanged outside Newgate gaol, in front of a crowd of 50,000 people.

His fate may have deterred others, for it was not until June 1881 that the next such outrage is recorded. A retired stockbroker, Mr Gold, was stabbed and thrown from the carriage in Balcombe tunnel on the Brighton-bound train. The suspect, Percy Mapleton, called himself Arthur Lefroy and pretended friends in high theatrical places, a conceit not unusual in that south coast resort, then or now. Like Muller, he was caught, tried and hanged, in Maidstone jail. Lurid drawings of the murder appeared in the Illustrated London News and the execution drew reporters from the local press. There were only three more passenger murders up to 1914, but salacious interest was fed by the story of a sexual attack by one Henry Nash, who forced himself on a lady traveller, Mary Anne Moody, on a Waterloo to Winchester train.

She jumped out of the carriage window onto the running board, and was clinging to the side when a passenger in the next compartment held her tight to the carriage. The alarm was raised by farmworkers in a field – there was no communication cord in those days – and Nash served nine months for his crime.

Rather more newsworthy was a similar assault on the same line, on Miss Kate Dickinson by none other than a friend of the Prince of Wales, Colonel Valentine Baker of the crack 10th Hussars. Once again, the victim escaped onto the running board until the train pulled into a station. The resultant trial created a public sensation, and the ungallant colonel was fined heavily and dismissed from the service on the personal insistence of Queen Victoria despite a plea for leniency by her wayward royal son. Baker spent his "bird" in some luxury, in his own cell, with visitors during the day bearing champagne and cigars. In the evenings, he wrote the story of his travels in central Asia. [A copy of which is on my shelves].

The reverse of this crime was not unknown: blackmail of a man by a woman passenger bringing a false claim of assault and attempted rape. One such, in 1866, involved charges brought by Ellen Allen against Alexander Moseley, a 29-year-old respectable married surgeon and father, alleged to have happened on a Watford to Euston train. But the Old Bailey court heard that the claimant was a former prostitute known at Victoria station for "molesting gentlemen". She was convicted of perjury and conspiracy, and jailed for five years.

These crimes and misdemeanours became the regular subject of musical-hall jokes, songs and paintings of the period. Questionable behaviour in the train also began to appear in literature and short film, *The Kiss in the Tunnel*, just over a minute long, made in 1899 by cameraman George Albert Smith with his wife Laura, shows them canoodling in their carriage, suggestively as they go through a tunnel. Sex on the train had arrived, officially, pre-dating sleeping-car lothario James Bond by the best part of a century.

Offences against the person make the most lurid headlines, but loot is the chief motivating factor of the career criminal and the railway brought unrivalled opportunities to steal high-value goods.

The first great train robbery took place as early as 1855, when a shipment of gold bullion bound for Paris from London Bridge was stolen on the night of 15 May by a four-man gang.

Professional burglar Edward Agar and his accomplice William Pierce, a former South Eastern Railway worker sacked for gambling, aided by two current SER employees, William Tester and James Burgess, used duplicate keys to raid the safe in the guard's van, taking three boxes of bullion and coins worth £12,000 (some £1.5million today). The theft was not detected until the shipment failed to reach the French capital and detectives were mystified about whether it took place in England or France.

It was only by chance, as is so often the way of these things, that the gang was caught. Pierce agreed to fund Agar's ex-girlfriend Fanny Kay after the gang boss was arrested on another offence, but ratted on his word. Fanny went to Newgate prison asking the governor for help and squealed on the robbers. They were all convicted: Pierce got two years' hard labour but the harshest sentence was reserved for the railway servants Tester and Burgess, who were sentenced to 14 years transportation to Australia. Agar, who confessed that he had not made an honest living since the age of 18, turned Queen's evidence and was not convicted of anything. He, too, was transported two years later.

And there the story ends, except that the robbery became a cause celebre that never failed to fascinate. A century later, the daring heist was made into a 1960 TV film starring Colin Blakely, and then a gripping comedy feature film *The First Great Train Robbery*, with Sean Connery playing Pierce as a kind of Raffles gentleman crook.

The excellent railway sequences were actually filmed in Ireland, at Dublin's Heuston station and on the Mullingar to Athlone line.

No record of railway crime would be complete without the modern Great Train Robbery of 8 August 1963, when the night Royal Mail from Glasgow to London was ambushed between Cheddington and Leighton Buzzard in Buckinghamshire. A gang of 15 men, led by Bruce Reynolds and including Buster Edwards, Charlie Wilson and Ronnie Biggs, aided by a renegade retired driver, attacked the train after it was halted by a fake danger signal. They got away with £2.6million (£70million today). Holed up in nearby Leatherslade Farm, they played Monopoly, leaving fingerprints that led to their eventual discovery and conviction. The ringleaders got 30 years. Train driver Jack Mills, beaten with an iron bar, suffered life-changing injuries and died of leukaemia in 1967. He has been remembered with name plates on two locomotives. Most of the money, in £1 and £5 notes, was never recovered.

The gang members went on the run to continental Europe,

Australia and South America before being brought to justice. The most famous fugitive, Ronnie Biggs, escaped from prison and led frustrated police a merry dance for 36 years, finally living in Brazil, from where he could not be extradited. He returned home voluntarily in 2001 and went straight back to jail, serving ten years of his 30-year sentence before being released on compassionate grounds aged 79. He died in a care home in 2013, having once said, "There is no honour in being known as a Great Train Robber. My life has been wasted."

It was not wasted by the small army TV and film producers, biographers, and fiction writers who have made a lucrative industry from the crime and its perpetrators. John Mills, son of the unfortunate driver of the night mail, said, "I deeply resent those, including Biggs, who have made money from my father's death." I share his view, which is why this, arguably the most famous crime on the railway gets less than one page here.

On the other side of this endeavour, the railway companies were not slow to set up their own police force. Minutes of the Liverpool and Manchester Railway in 1830 refer to "The Police Establishment", probably men sworn in under a law of 1673 to protect the construction of the line and control traffic. Station houses were provided at one-mile intervals, and this is where the term "police station" is likely to have originated. An Act of 1831 gave them wider powers, and constables carried truncheons elaborately painted with the crest of their company, which are museum pieces today.

The arrival of thousands of navigators – the "navvies" who

built the lines – put a heavy workload on the nascent force, and a further Act of 1838 placed financial responsibility for keeping the peace in shanty towns springing up by the construction sites on the railway companies. It was no easy task. The following year, a fight between English and Irish navvies building the Chester and Birkenhead line lasted four days before order was restored by a detachment of infantry.

In 1840, two labourers murdered ganger – or foreman – John Green on the Edinburgh to Glasgow Railway and soldiers were again required to assist. Justice must have been swift, for the perpetrators Dennis Doolan and Patrick Redding were hanged on a makeshift scaffold between the tracks.

But the lure was chiefly loot, as previously observed. The railway began carrying mail in 1838 and thefts began shortly afterwards. On a single day in 1848, the Eastern Counties lost 76 pieces of luggage and in the following year, theft cost the six largest train companies £100,000. The courts were particularly severe on company servants: in 1873, ten railmen were each sentenced to ten years' imprisonment for stealing from their employers.

The 20th century brought radical change. With over half its members conscripted for the Great War, the Great Eastern Railway recruited nine women as constables, one of the first to do so. The North Eastern introduced dog patrols. When the 1921 Railways Act created the Big Four, each had its own police force controlled by a chief of police.

In the 1926 General Strike, railway police issued identity cards to volunteers. In World War Two, women were again recruited, and this time they kept their jobs. The force was doubled in size to meet the growing threat of stealing, made easier by the blackout and in post-war years, more attractive by rationing.

Between 1941 and 1952, thefts on the railways exceeded the total number of all others reported to the combined police forces of England and Wales.

After nationalisation, the forces of the Big Four were consolidated into the British Transport Police, with a membership of 3,890. It is still responsible for crime prevention and passenger security today. The campaign message "See It, Say It, Sorted" together with the text number 61016 is heard on every journey, sometimes with greater frequency than some travellers might wish, to which no doubt the reply is "you can't be too careful". And while it is true that rail is the safest form of travel in Britain, with only 16 crimes recorded for every million journeys, there is still risk.

In 2023, almost 4,000 violent assaults were committed on the network, 1,419 of a sexual nature and 699 robberies. The figures were obtained by a Freedom of Information request by the Liberal Democrats, who claimed that train carriages were being turned into crime scenes with criminals allowed to act with impunity. BTP has conceded a "notable increase" in crime, and in 2024 recorded 9,376 crimes at the ten worst-hit stations. 3,831 serious, including sex attacks and robberies. Manchester Piccadilly was the worst offender and subterranean Birmingham New Street second.

London Bridge was in third place. The railway is not even safe for staff, with nine out of ten saying they have experienced violence at work.

But rail is still safer than road, and infinitely more comfortable. And while some stations are not exactly home from
home, they are better than airports. The trouble is, travellers often have no choice, which was really the point of the
railway in the first place.

CHAPTER 24

THE ONE THEY THREW AWAY

IF A VOLCANIC eruption of the kind that engulfed the Roman city of Pompeii in 79AD covered parts of this country, future archaeologists would puzzle over some of the mighty edifices they excavated. Great bridges, deep tunnels, strange embankments and steep-sided cuttings. These massive structures must have played an important role in an earlier civilisation, but what was it? And which ancient tribe built them?

Why did a 1,527ft-long viaduct with 21 arches and a lattice bridge stride across the River Don in Conisbrough, South Yorkshire, at a height of 116ft? The answer lies in the railway culture. This goliath of girders was opened in 1909 by the over-optimistic Hull and Barnsley Railway, but abandoned in 1966, a preposterously short life of less than 60 years. It stands today as a monument to the folie de grandeur of this long-forgotten minor railway company.

If we turn our gaze to the ground, and below, we find

around 600 tunnels left behind by the railway, mostly derelict and unused, given back to nature and geology, mindless of the heroic struggles of the men who hacked them from solid rock. The longest exceeds three miles, at Woodhead in the Pennines, where electrified lines bored 5,346 yards to connect coalfields on the east side with power stations in Lancashire. Opened in 1954, it closed in 1981, alongside an earlier bore of almost the same length opened in 1854 and closed in 1954 when the new tunnel came into use.

This is the railway they threw away, infrastructure discarded on a grand scale that would furnish material for a dozen TV abandoned engineering programmes. Some of it you can walk over, or under, or through. There are websites for aficionados like Forgotten Relics, and new attractions – if that's the right word – are being added all the time. Some of the tunnels have flooded, others are partially buried. Others are fenced off with high gates and they can be very unsafe places.

But they always were. The railway was nothing if not daring in taking on nature. Take the wrought-iron latticework Bennerley Viaduct, nicknamed the "Iron Giant" opened in 1878 by the Midland Railway to connect Nottingham and Derby Friargate, a station long closed. The World Monuments Fund calls it "extraordinary" and Historic England "a stunning example of the genius of British engineering". Yet it survives only because it would cost too much to demolish.

This remarkable Grade II listed structure, one of only two of its type in Britain, is 1,452 yards long on 19 spans, and crosses the river Erewash valley at a height of 60ft. Its unusual construction, thought to be modelled on the Viaduc de Busseau, opened in 1864 in Central France, is due to severe geological problems caused by extensive mine working in the valley.

Bennerley survived a Zeppelin bomb attack in World War One, when nine airships conducted what was known as the Great Midlands Raid on 31 January 1916. One dropped seven high-explosive bombs in the vicinity, causing minor damage to the viaduct. Shrapnel marks can be seen on one of the piers. The site was abandoned by British Rail in 1968 and demolition was deemed too costly, so it was left to rust until campaigners fighting to keep their favourite "giant" won £1.7million for funding its refurbishment. It has now reopened to walkers and cyclists, a tribute to local determination every bit as strong as the ambitions of the Victorian engineers.

This is also one of the few engineering miracles that finds a place in English literature. DH Lawrence, born in nearby Eastwood, Nottinghamshire, mentions it in his novels, most noticeably in the classic *Sons and Lovers*, "There was a faint rattling noise. Away to the right, the train, like a luminous caterpillar, was threading across the night. The rattling ceased. 'She's over the viaduct. You'll just do it'."

More recently, after a visit during restoration, Matthew Parris, former Tory MP in the region, wrote in his column for *The Times*, "It is, on one view, a hideous thing, and on another a precious and remarkable monument to early railway engineering." I do not have a bucket list of things to do before I die, but if I did, walking across this magnificent viaduct, built to unique specifications to overcome the dangers caused by mining, would be slopping on the brim.

These amazing structures are also a favourite of film makers. The breathtaking Thornton viaduct, a 900ft-long S-shaped behemoth soaring 120 feet over Thornton Beck in the Pinch Beck Pennine valley on the old Great Northern Railway, appeared in an episode of BBC comedy *Last of the Summer Wine*.

This improbable adventure – weren't they all – involved the three old men and a mangle. Interrupted by two policemen trying to hoist the ancient mangle up over the parapet, Seymour, Compo and Clegg drop it on the police car, to great canned laughter. It is a favourite episode for *Summer Wine* lovers. The viaduct opened in 1878 and closed to traffic in 1966. Now open to walkers, it carried the line from Bradford to the mill town of Keighley, today's home of the heritage line on which the first version of *The Railway Children* was filmed in 1970.

Mytholmes tunnel on the Keighley and Worth Valley Railway where Jenny Agutter waved her red flannel petticoat to stop a train is most definitely not open to the public. But according to the website Forgotten Relics Of An Enterprising Age, you can walk through quite a lot of the most historic railway holes in the ground. Take Combe Down tunnel, on the old Somerset and Dorset, 1,829 yards long, opened in the southern suburbs of Bath in 1874 and closed in 1966. It now forms part of the Two Tunnel Greenway unveiled in 2013 as a shared-use cycle and footpath.

Walkers may not be aware of this tunnel's grim history. As the longest unventilated bore of its time, Combe Down was always a sore trial for footplatemen. On 29 November 1929, the driver and fireman of S&DJ 2-8-0 number 89, running tender-first at the head of an overloaded northbound goods, were overcome by smoke. The train ran away, crashing into a goods yard outside Green Park station, killing the driver Henry Jennings and two railway workers in the depot.

The story of the railway is littered with such tragedies. Victorian engineers looked on the deep valleys and high highs of the British countryside as obstacles, a challenge to be overcome. Great was the loss of life in realising that ambition. In 1846, the Huddersfield and Manchester Railway began a single-track tunnel under the Pennines, adjacent to a canal tunnel built some years previously, access from which greatly assisted the work. 1,953 navvies working on 36 faces drove the bore 85 yards a week. Construction of the three mile 57 yard link took two years, during which nine men died. The project proved insufficient for the volume of traffic and a second was built in 1871. That, too, proved inadequate and a third, twin-track bore was added in 1894, using 1,800 workers billeted in a paper mill at Diggle on the eastern side. The original single-bore tunnels were taken out of use in 1966/70 and are closed to the public.

Today, the 1894 version carries Transpennine services. I have frequently taken the train through this cavernous railway burrow, sometimes pondering the human cost of my comfort. The canal tunnel is now open for tourist trips but I don't think I will be in the queue for a seat on the "discovery boat". Not at £50 a pop at the time of writing.

Anyway, the best way to explore our lost railway heritage is by Shanks' pony: on foot. Mile upon mile of former track – permanent way that proved all too temporary – is open to walkers, from remote Speyside in the Highlands to the Plym Valley on the edge of Dartmoor, the Brecon Beacons to the Isle of Wight. This is an increasingly popular pastime and there are guide books galore to show you the way. The best, Julian Holland's magnificently-illustrated, coffee-table-size *Exploring Britain's Lost Railways*, also comes in a handy portable version.

I took it with me on a walk with my late friend Robert, from Low Moor to Dewsbury on the Spen Valley Greenway. This is a seven-mile hike from the south of Bradford, on the trackbed of the former Lancashire and Yorkshire line to Huddersfield through towns with impossibly-Tyke names like Cleckheaton, Liversedge and Heckmondwike. Comics invented the satirical portmanteau name "Heckmondsedge". This was "shoddy country" where cloth was made from old woollen clothes, though there was nothing shoddy about the finished product. Running through the textile mills, the line opened in the mid-19th century and thrived until the industry it served began to dwindle after World War Two. It even had a direct train on old Great Central Railway metals from Marylebone, the South Yorkshireman. Passenger services on the southern section to Thornhill, stopping at Ravensthorpe, ended in 1962, even before the infamous Beeching Axe descended. They continued on the section to Mirfield for another 23 years before closing and freight lasted on the leg to Heckmondwike until 1981. The whole line shut in 1990 and far-sighted civic leaders reopened the trackbed as a footpath and cycleway, a green corridor through dull country enlivened by a series of artworks.

Low Moor was distantly familiar to me as the location of a large motive power depot in the 1950s, home largely to goods engines and some powerful 2-6-4 tanks for passenger services in the hills. The 12-road shed lasted until almost the end of steam, finally closing in 1967. My Holland bible tells me there were also extensive carriage sidings, but I don't remember those. It once boasted a triangular junction, but I don't remember that either. Low Moor is now a commuter stop on the line from Bradford to Halifax, the first stop out of "Wool City" which is this year's City of Culture.

It wasn't then, on my departure from what is now known as Interchange station, a horrid little two-platform affair linked to the even less attractive bus station alongside. But the sun was shining and Robert, who'd come from Huddersfield, was waiting. This was a Saturday morning jaunt, not a serious walkathon. It was before my "event" (an aortic dissection, from which I was brought back to life by surgeons at Leeds General Infirmary) so I didn't then walk with a stick. Even so, no need to knock yourself out. We ambled and talked and ambled and talked, as the way two old hacks do, and I faintly recall seeing the artworks mentioned in the guide: Sally Matthews' flock of Swaledale sheep made from scrap industrial metal (I see the real ones every day), Alan Evans' giant pedal and cycle set and Trudi Entwistle's

40 giant steel hoops set in a circle. But I cannot claim they made a big impression and besides we were getting a bit thirsty. I'm sure we made it past Cleckheaton, where the station site is now a supermarket, but a mile before the end of the greenway we turned into Liversedge for a pint. Once off the track, so to speak, there didn't seem much point in going back, so we took a bus into Bradford to continue the refreshment at the Sam Smith's pub next to Forster Square station where I could get my train back to Skipton.

Thinking back, it feels like a poor show. I didn't take any notes – that would have turned a Saturday stroll into work – but I have the Holland guide to refresh my memory. He lists more than 50 walks and would that I could do them all. Or just a handful: maybe Bewdley to Tenbury Wells, where my fireman's memories book has already taken me, or the Dunblane (where I covered the shooting tragedy for *The Observer*) to Crianlarich. I did "do" the full length of the former Leek and Manifold light railway – all eight miles from Waterhouses to Hulme End and indeed on to Harting-ton in Derbyshire to catch the first available bus. That was a long time ago, back in the eighties, before I was in mine.

And I did venture on the Morecambe to Lancaster trail when Mrs R and I had a weekend at the Midland Hotel.

But that's about the sum of my pedestrian peregrinations. I can, and do, still read the guide and ponder. If you are able to follow their lead, you will be assured of days in the open air, with very few inclines and no hills, amid scenery that still feels like "here was once a railway". Just beware of the crazy cyclists.

CHAPTER 25

THE ANTIQUE WORLD OF THE RAILWAY

"NO MORE WILL I go to Blandford Forum and Mortehoe, On the slow train from Midsomer Norton and Mumby Road," sang the comedy duo Flanders and Swann.

Only months after Dr Beeching published his infamous report in 1963, they were lamenting the likely closure of the much-loved Somerset and Dorset Joint railway line.

Opened in 1854, this largely-rural line meandered through the Mendips from Burnham-on-Sea through Bath to Poole, serving no large centres of population and very little purpose except peak holiday traffic and traffic from the Somerset mines.

The coalfield, in part owned by the Rees-Mogg family, became exhausted and holiday patterns changed. People from the Midlands conurbations flocked to the Costa del Sol, not the beach, but Bournemouth on the Pines Express, hauled by a steam locomotive. Various mocked as "the Slow and Dirty" and "the Slow and Doubtful", it finally closed on 6 March 1966 after years of campaigning by local people and enthusiasts failed to dissuade a Labour government that came to power promising no more major closures.

But you can actually go from Midsomer Norton today, though not very far and certainly not to Mortehoe because the S&DJ never did either. Enthusiasts have restored a mile-long section of track from Midsomer Norton south towards Chilcompton, one of the original 48 stations on the line. Trains run at weekends in the heritage season.

This is a modest enterprise, but it shows what can be done with sufficient determination and it's just one of more than 130 restorations that have brought the railway back to life in places where it succumbed to the Beeching Axe and later closures. These heritage operations attract 13 million visitors every year, generating an estimated £600million-plus to the economy. The pioneers of the earlier S&D – Stockton and Darlington – would be proud of them.

I haven't been to all the preserved lines, but I have been to most of them. I have even driven, alright, travelled on the footplate of, one: the most famous engine of its type still running, 60007 LNER Pacific Sir Nigel Gresley, named after the great designer himself. This adventure took place on the North Yorkshire Moors Railway, one of the oldest and best-run heritage lines, running for 18 miles through some of the most remote, beautiful Yorkshire countryside, plus a further six miles on Network Rail metals into Whitby. Built by George Stephenson and opened in 1835, it was a boon to local farmers, but the Beeching Axe descended in 1965 for passengers, and freight a year later. Enthusiasts began campaigning for its restoration within months and their efforts have yielded the country's most popular heritage line.

I've visited the NYMR many times and even holidayed in a cottage so close to the line I could also touch the great hissing monsters of delight. My account of life on the footplate appeared, appropriately enough in the journal of the train drivers' union Aslef, in October 2013, headlined 'You Can Take The Boy Out Of Railway Terrace But You Can't Take Railway Terrace Out Of The Boy'. It reads: "I know it always took a long time to get into the top link on the footplate, but this is ridiculous. It's taken more than half a century since I envied train drivers down our street to get my turn on the regulator of 60007 Sir Nigel Gresley, thundering up a 1:49 on the North Yorkshire Moors Railway.

"There's an adrenaline rush on the bucking, heaving footplate, the fire like a gaping mouth of Hell, roasting and blinding. So this is what it's all about! No wonder there was such great comradeship in the steam days.

"Thankfully, for the safety of the travelling public, I'm not allowed more than a touch on the magical lever of power. Puffer nutter I may be. Qualified, I'm not. Today's driver of this magnificent thoroughbred, the 100th Pacific built at The Plant in Doncaster, is a 28-year-old accountant from Filey, Rob Green, who doesn't look old enough to clean this A4, much less take her out on the road. He's one of a new breed of newly-trained footplatemen on the heritage system – there are more than 100 such lines round the country – and his fireman is veteran Gary Stainburn, 45, an IT manager from Castleford. Most of the shovelling on this 11-mile run is done by Jack Prince, a 17-year-old cleaner on his first firing turn with 60007.

"Gary has a wealth of experience, having driven all three surviving steamable A4s (the others being 60009 Union of South Africa and 60019 Bittern) including stretches on the East Coast Main Line. He is almost poetic about his engine. 'She sounds like a sewing machine on the main line,' he sighs. 'She was built to do 100mph for seven hours at a stretch, rather than 25mph.' That's all we're allowed to do on this line. Our seven coach train is overpowered and the effect is overpowering. Crowds come out to see 60007 blasting up the incline to Newton Dale, to the 532-ft summit at Fen Bog, where the line's constructor Robert Stephenson 'floated' the permanent way on wattle fences over heather-stuffed fleeces.

"Driver Green, introduced to steam railways in a carry cot by his rail enthusiast father, has only to tug the regulator to feel the full force of Gresley's masterpiece. 'I came here on family holidays,' he tells me, 'now I sit here and do it and I have to pinch myself that I've actually achieved it.' There is no top link on this railway, but there is a Whitby link of drivers passed to take trains on to Network Rail, to the premier resort on Yorkshire's Riviera Coast to the railway's own platform.

"Some scoff at these 'weekend footplatemen', but the NYMR is a thriving business, carrying 350,000 passengers a year, infinitely more than in the days of British Rail. It employs just over 100 full and part-time staff and contributes tens of millions to the local economy."

The railway has appeared in many films, and practically has a speaking part in *Heartbeat*, the long-running and hugely-successful TV series about police officers in "Herriot Country". This is my favourite, but, as they say, there are others. So let's go on a journey of discovery of these magical attractions, starting in Scotland, where the Strathspey Railway runs through the hills of Cromdale from ski resort Aviemore to Broomhill, via Boat of Garten on a line closed in 1965 and reopened by enthusiasts. This magnificent journey, on the former Highland Railway route to Forres, climbs through heather moor with views of the Cairngorm mountains, through farmland to "Boat" by the river for ten miles to Broomhill, a replica station built on the foundations of the original.

There are plans to extend the line to Grantown on Spey, the next town in the valley, but these are on hold because the cost of diverting a main road and building a new bridge has risen dramatically to more than £13million.

Scotland has no fewer than eight preserved railways, but this is nothing compared to Wales, which boasts 18. However, all but four are narrow gauge. This is the land of "the great little trains".

The great little nation does have standard-gauge lines at Blaenavon in the south Wales Valleys, and a burgeoning enterprise at Gwili in Carmarthenshire, but the jewel in the crown has to be the ten-mile Llangollen railway in Denbighshire. This line, on part of the old Ruabon-Barmouth route opened in 1862 and closed in 1965, came back into life in 1975 thanks to volunteers who refused to see it die. Despite financial problems in recent years, it's now a flourishing concern operating frequent trains through the glorious Welsh countryside from the market town of Lalngollen to a new station at Corwen. There is talk of an extension to Ruabon, but that's all it is so far.

In England,where to begin? Still in Yorkshire, we have the underrated Embsay and Bolton Abbey Steam Railway, formerly on the Skipton to Ilkley line, a four mile jaunt through the Dales almost to the site of one of the country's most beautiful abbey ruins. But let's not be too patriotic. Lancashire may not have much to offer, but Cumbria takes you right into the Lake District on the steeply-graded three-and-a-half-mile former Furness line from Haverthwaite, near Ulverston, to Lakeside, on the shore of Lake Windermere where pleasure boats ply across to Bowness.

Moving south, the Great Central Railway on the old main line from Sheffield to Marylebone "deserves a detour" as the guide books say. Based at Loughborough, it brags of being the only main line heritage railway and it certainly lives up to the boast on its eight-mile, double-track run to Leicester North via Quorn & Woodhouse and Rothley.

This route, the last built into London in the late 1890s, was the biggest single casualty of the Beeching closure programme. It would never have been closed today and would arguably have been a wiser – and cheaper – route for HS2. The GCR has featured in more than a dozen films, including *Enigma* and TV versions of *The 39 Steps* and *The*

4.50 from Paddington – showing that it's the only preserved line where you can see steam trains passing each other at speed. The line is being extended north back to Nottingham, creating the longest heritage route and linking two university centres.

I have a soft spot for the Didcot GWR centre, mentioned elsewhere and I find it hard to pick a favourite in Great Western Territory. The Dartmouth, running through the most attractive seaside locations from Paignton, Devon, is beloved of railway photographers. The Swindon and Cricklade, not easy to find, is worth the effort. The South Devon, seven miles of lush scenery along the river Dart, from Totnes to Buckfastleigh, recreates an idyllic branch line of yesteryear. But the palm has to go to the West Somerset: the British Rail of my generation reborn. The longest of the "heritages" it winds for 20 miles from the coast at Minehead to Bishops Lydeard on the outskirts of Taunton, with occasional specials through to the county town. En route, the line, closed in 1971 and progressively reopened five years later, passes through – if you can fight the temptation to get off – Watchet and Blue Harbour, with superbly-maintained GWR locos at the head of the train. Unmissable for puffer nutters like me.

Space limits my enthusiasm and I haven't even touched on the opportunities for volunteering – be your own station porter! – or the engine-driving courses offered on the best preserved railways, but I really must mention the narrow-gauge sector. It's not my go-to, but is the abiding passion for many others. "Small is beautiful," urged thinker EF

Schumacher in his groundbreaking book of 1973. He was talking about economic policy, not railways, but his maxim has always been close to the heart of lovers of "the great little trains". And, astonishingly, there are more narrow-gauge enthusiast operations than standard-gauge heritage lines in Britain today, ranging from not much bigger than toys in the garden to the real-life-in-miniature Romney, Hythe and Dymchurch in Kent. More are coming all the time, like the magical reconstruction of the cliff-top Lynton and Barnstaple in Devon and the Bala Lake railway in central Wales. The appetite for doing it differently is unabated. More to come on them later.

The future for railway revival is bright indeed. Halted by the pandemic, heritage lines are once again thriving commercial ventures, giving employment, generating huge economic returns and attracting millions of people keen to get out into the countryside. They are in the forefront of celebrations for Railway 200, with dozens of events taking a full page in the excellent *Heritage Railway* monthly glossy magazine. Blow that whistle, raise that green flag, all aboard!

Apart from our incomparable preserved lines, Britain also has a wide variety of heritage that will never turn a wheel again, in dedicated museums. Most nations with a railway history have a railway museum, often tucked away on a barely-accessible site, as if they were ashamed of their past. Nowhere is this more true than railway-mad India, where the Delhi museum was full of rusting hulks last time I managed to find it, via a moped taxi whose driver claimed never to have heard of it.

Some are just graveyards for old locomotives that the authorities evidently don't know what to do with, like the steam cemetery I encountered in Kalamata, Greece. Others, like in Zagreb, Croatia, are well-run and offer a glimpse into the history of transport. But the greatest of them all is Britain's National Railway Museum in York, telling the glorious story of the nation's greatest invention and visited by 700,000 people every year.

Housed in the city's former motive power depot – 50A to the cognoscenti – it has the world's largest collection of historic locomotives, carriages, waggons, signals and other infrastructure, plus the most extensive library, the largest hoard of railwayana from nameplates to hotel crockery. The only thing it doesn't boast is a bar, despite the presence of some very fine dining cars (retired).

The origins of collecting "railwayana", as the trade in the industry's vast range of antiques has become known, is lost in the mists of the steam age. Enthusiasts have always sought out historic artefacts to keep the flame of their passion alive: station signs, signals lamps and suchlike. But for my money, as good a place to start as any is the water-tank turret of Doncaster Grammar School in 1936, when inspired school-boys Tony Peart and Wilton Jones began accumulating name and number plates from locomotives scrapped at "The Plant" rail engineering works in the town (now city).

Their intention was to build a Museum in the Tower and over the years they and their successors, including masters drawn into the enterprise, built up an astonishing hoard of 10,000 exhibits, the largest in private hands. This cornucopia was displayed on walls and the floor inside the turret, accessible to limited numbers of visitors on special occasions, by a winding iron staircase. More than 80 years would pass before they were put on magnificent display in Doncaster Council's new-build Danum Gallery Library and Museum. This Valhalla of the railway is now a must-visit for rail enthusiasts.

CHAPTER 26

RAILWAYS
OVERSEAS

A 19TH CENTURY SAYING, "trade follows the flag", described the commercial exploitation of colonies by European nations anxious to acquire an Empire. And where British ships planted the Union flag across the globe, the railway followed very soon after. It may be said that trade followed the red flag first waved 200 years ago, ahead of the Stockton and Darlington service of 25 September 1825.

The speed and comprehensive nature of this export drive cannot be overestimated. The wealth of India, Africa and South America was exploited for more than 100 years, with a vast network of lines, often through the most challenging territory and with little thought to the impact on local people. A great many of these railways still operate and at least one remains in British hands.

That is the high-altitude route from Antofagasta in Chile over the Andes to La Paz, capital of Bolivia, which crosses the barren, baking Altiplano at a height of more than

14,000ft. With the late Tom McGhie, of the *Daily Star* (I was then chief reporter of *The Observer*), I took this journey in December 1986. It is supposed to take 27 hours. It took 49, after an ambush by bandits. More on this later. First, let's look at the English pioneers who had their sights set on somewhere closer to home: Paris.

William Mackenzie, a native of Nelson, Lancashire, an engineer who started life working on canals, and Thomas Brassey, a Cheshire-born civil engineering contractor, joined forces to take the train to the continent. They built the first French railway, from the capital to St Germain, a distance of 20 miles, in 1837, only 12 years after the trailblazing Quakers of Darlington began the world's first passenger carrying service.

British investment poured in thereafter, financing lines to Rouen, Le Havre, Orleans, Tours and other provincial cities including Boulogne. Indeed, gourmand Mackenzie became so enamoured of France that he went to live in Paris, where he was an habitue of the best restaurants.

The British brought unrivalled construction skills and equipment to continental Europe, and also the men who built them. Up to 5,000 navvies worked on the Rouen line alone, where it was noted that they "enjoyed the abundance of cheap wine". They were also paid up to twice as much as French labourers and given their occasionally rumbustious behaviour on weak beer in the English shires, the impact of this largesse may be imagined.

Brassey, a farmer's son who became the greatest builder of his day, constructed not only a third of British railways but three quarters of the lines in France and the rest of Europe, plus ventures in Canada, Australia, South America and India.

It was said that at the time of his death in 1870, aged 65, he was responsible for one in every 20 miles of line in the world. Amid this prodigious output, he also built docks, bridges, viaducts, stations, tunnels and part of Joseph Bazalgette's London sewerage system still working today. He is commemorated by a bust in Chester cathedral and a statue outside the city's station.

Other British pioneers joined the rush to rail the world. George Stephenson, the acknowledged Father of the Railway, engaged in projects in Belgium and Spain, while his son Robert also worked on railways in countries as diverse as Norway, France and Egypt. His famous works in Newcastle exported engines, opening up a private sector market that was still going strong well into the 1960s, concentrated in cities such as Leeds, Glasgow and Manchester.

One controversial example of Britain "taking over" a nation may be found in 19th century Argentina, a country covering more than a million square miles – 11 times the size of the UK. After gaining independence from Spain in 1816 and years of civil wars, the country finally took to the iron road in 1857, with the first line from the capital Buenos Aires eventually to the up-river port of Rosario, to serve the vast wheat and cattle-producing farmlands of the interior.

This first line on the Ferrocarril Oeste de Buenos Aires was built by William Bragge. Birmingham-born civil engineer Bragge was Chief Mechanical Engineer with the Birkenhead

Railway before heading to Brazil to survey the first railway there and then to Argentina. During all this South American travel he managed to compile a 17-volume library on the literature of tobacco.

British finance and know-how promoted construction rail in Argentina, with many more lines reaching across the pampas to the deep south and west to the Andean border with Chile. The terrain was suitable, but work was hampered by outbreaks of malaria, cholera and typhoid fever and attacks from indigenous tribes who hated and feared the coming of the train. In one recorded instance, they even tried to lasso a locomotive.

Eventually, the system grew to 29,000 miles by 1930 – the largest outside Europe, and 69% British-owned. This extraordinary expansion allowed mass migration, chiefly from Spain and Italy, to more than double the population to four-million-plus by the end of the 19th century. As the industry's official historian, HR Stones, observes, "It was the first time in the history of railway construction in a foreign country that the promoters became colonisers without being preceded by military conquest."

William Bragge was followed by another English pioneer: Edward Lumb, a merchant adventurer, built the British-financed Great Southern Railway to a 5ft 6in gauge in the 1860s. A prominent member of the Buenos Aires business elite, he is also remembered for playing host to Charles Darwin returning from his groundbreaking scientific expedition on the Beagle in 1833. Such are the ways in which the railway and progress are intertwined.

Having played such a huge role in the development – some would say, exploitation – of this vast country, it was predictable that the system would one day fall into the hands of the Argentinian government. Nationalisation came in 1948, at a price paid to the British owners totalling £150million. President Juan Peron was undergoing an appendectomy and could not attend the signing ceremony "amid widespread celebrations" so his place was taken by his wife Eva.

The event does not figure in the hit musical of her life, *Evita*, which is a pity because it was accompanied by a hoisting of the national flag and a prolonged blast on the whistle of the veteran locomotive La Portena by the country's oldest engine driver, Pascual Corera, though admittedly it might have been difficult to stage at the 1978 premiere in London's Prince Edward Theatre.

In the years that followed, the Argentine railways went through a demoralising cycle of decline, closures and rena-tionalisation, as if the British were still in charge. Most of the important cities are served by two state-owned companies and there is even talk of a Chinese-built high-speed line from the capital to Cordoba. Few now recall the Victorian builders from 7,000 miles away, though one of the local stations bears the iconic name of Lumb.

With its staggering, construction-defying Andean range, South America has always been high on the puffer nutter's bucket list. As mentioned earlier, I succumbed to the mystique in December 1986, after learning that the Antof-agasta-Bolivia railway was still British owned. I interviewed an executive of the company in the city, and secured a laisser-passer for the journey.

This is a journey of a lifetime, that can probably never be repeated, on Ferrocarril de Antofagasta a Bolivia, west to east from the Chilean hinterland to La Paz, capital of Bolivia – 715 miles on a narrow-gauge line, into the clouds and through breathtaking (literally) scenery. Like the Stockton and Darlington, this was originally built to carry goods, not people, in this case rich deposits of nitrates and copper. Construction started in 1873, with mules providing motive power. Steam engines appeared three years later.

The FCAB must be one of the few railways to have caused a war: the so-called War of the Pacific, between Bolivia, in which Antofagasta was then situated, and Chile between 1876 and 1883, partly caused by the Bolivian government's demand for back taxes from the railway. Bolivia lost and the western end of the line then became, and remains, in Chile. Originally with different gauges either side of the border at Oruro, the whole length was eventually metre-gauged by 1928. It once proudly boasted The International train, with sleeping and dining carriages.

This was our objective, but the 1980s version was, shall we say, a little downgraded. But it was worth the long haul. Flying with Journey Latin America to Santiago, McGhie (a seasoned traveller, his diplomat father was once Our Man in Saigon) and I travelled by bus up the Pacific coast to Antofagasta.

The weekly Wednesday train, an official informed us, would take 27 hours, "unless there is a strike in Bolivia". We were advised to buy food and water for our transit across the desert, where rain never falls. Before setting off, we inspected the company museum, a treasure trove of railway-ana, including old photographs of desperadoes who preyed on the line, the Winchester repeater rifles of the guards, and of the old steam locomotives mainly built by Hunslet and Kitson, both of Leeds.

These days, the western terminus of the line is at the desert town of Calama, some 145 miles inland. Here, we travelled in an FCAB minibus with other travellers, one of whom had fish in a shampoo bottle and another with chickens in a cardboard box. The bizarre started early. Some carried blankets, I noticed with dismay. We had none, for the tiny private "compartimento" of our own, with leather uphol-stered chairs and beds that let down on ingenious brass chains like a drawbridge: metal springs, but no bedding. The temperature at night falls as low as -15 °C.

As my diary of the trip records, nothing is ever what it seems in Latin America. Being allocated a private accommo-dation and taking possession are quite different things. Two very large bearded young men, shifty-looking and dragging vast bundles, demanded our cabin. "Interpol," they hissed, and we presumed they were on the run but they produced heavy metal badges and plastic identity cards and demanded more vigorously.

"I don't care if you're bloody Hercule Poirot, you're not having these seats," I expostulated, perhaps unwisely. A harassed official found them somewhere else – but not before they frisked our passports to show who's boss. You don't mess with Chilean police too much. Armed carabinieri strutted around the station in threes and even the railway staff carried business-like nightsticks. At 12.57, two minutes late, we pulled out: not even General Pinochet, Chile's military dictator since 1973 could make the trains run on time, particularly an English one.

Heat shimmered off the platform as we set off with a hoot from the big orange General Motors diesel loco (steam ended in the sixties) on the great adventure. Children waved in the suburbs of Calama [pop. 60,000, "city of sun and copper"], while to the north the giant smokestacks of the Chuquicamata mine belched into the thin desert air. This is superlative country: the highest railway, the direst desert and so on, so naturally "Chuqui" is the world's largest open pit copper mine, a 10,000-ft hole in the ground. Its spoil heaps are so vast they create a mini, man-made Andes for several square miles.

Inside the closed world of the train, travellers settle down as if for a siege among their impedimenta. They talk rapidly, but not to these gringo.

"They think you are American," says an Italian antique dealer (or so he says), the only other European on the train. "They don't like Americans." Admiring our accommodation – "Is very good!" – he stows his huge travelling bag with us. I make a mental note to deny everything should the boys from Interpol find any contraband.

Outside, the scene is practically unchanging, mile after mile of arid, treeless steppe, giving way to rounded hills the colour of old putty. The train lumbers on, sometimes at walking pace, sometimes at a steady jog, for the most part in a dead straight line. At San Salvador, 2,483 metres altitude, an hour later, nobody boards, and two women who got off to fill water bottles are almost left behind, but they could have gathered up their voluminous skirts and chased us down. At Cere, 2,841 metres, a man in a battered straw hat waves languidly. The permanent way gangs, perspiring under tin hats, are more animated, gesticulating and shouting. This once-a-week event must be a treat.

Three hours into the journey, our train is trundling steadily downhill, through a rock-strewn depression, with a limitless salt pan stretching away to the south. The heat would be intolerable, but I propped the window open with the hardback *Anna Karenina* that has been my constant companion since leaving London. A light breeze plays past its buckling spine.

Conchi, 3,146 metres, with its deep blue reservoir fed by the Rio Loa, a river here, but something we might call a beck in the Pennines, is next. By my, admittedly shaky, arithmetic the line has now passed the 10,000ft level and the air is noticeably thinner. Nobody smokes. Chests heave, seeking out the oxygen. Nausea in the stomach. Ears ache. The altitude induces light-headedness, as though someone has removed your brain and replaced it with cotton wool. Sounds like tolling bells reach my ears, but it's only the wheel of the ancient carriages straining on a tight curve. I notice tin cans and bleached skulls (llama?) by the lineside. We can't be going more than five miles an hour.

On and on, grinding the miles away, until we reach the

summit at Ascotan, elevation 3,956 metres. An official knocks before entering our thick compartimento door, he stinks of perfume. The speed of this train is ridiculous, I note with a trembling hand. We dined at 18.00, a disaster. Our emergency rations of cheese biscuits Tom bought have holes in them and anyway the pate has melted. Meal over in four minutes. We fantasize about food. Tom wants a leg of lamb with roast taties and cauliflower. I would happily settle for a bacon sandwich. It's a question of upbringing, I suppose.

Meanwhile the train goes at walking pace round a perfect volcanic plug, but the romance of FCAB is wearing a bit thin. After 20.00, the wind gets chillier, the sun is dipping lower, the great massif approaches and we do our best to sleep, fitfully. I wake at 01.20 on 11 December – my 43rd birthday – at the Chile-Bolivia border town of Ollague, where the coaches for La Paz are on an adjoining line. A mad scramble ensues, with little old ladies showing remarkable vigour and stylish elbow-play, but we get our seats 13 and 14 on coach 1401, the better of the two. Ollague, 3,695 metres, in the dark is a shadowy series of buildings, most prominently a police station.

It is 03.06 when we set off again, the coach a mountain of luggage. We travel only a mile or so when the engine is detached and another one takes over. All this performed higher than 11,000ft. A woman complains of "altura" sickness and groans. People laugh and then she cries. They give her aspirin and she regains her composure. I count: she has 12 bags. Belatedly, I realise: this is the Smugglers Express.

At 04.40, Avaroa, 3,701 metres, announces the Republic of Bolivia. Officials take away our passports in a blue zipper bag. I wonder if we will ever see them again. It takes three and a half hours to cross the border. A posse of soldiers, including a surly brute with a face like a startled mule, check the bags. By 06.40, the sun is coming up and we are deep into Bolivia, stopping at the remote settlement of Julaca, 3,655 metres, then Rio Grande and the large town of Uyuni, where dozens of steam locomotives rust in the sharp sunshine, row upon row in a large works. More intricate shunting and we bought cheese and bread after changing the last Chilean pesos.

It is past 10.00 by the time we leave and the mystery of the lady opposite's big smelly cardboard box is solved. "Cantitas! Cantitas!" she says. Singing birds. She is smuggling budgerigars! Chita is passed at 11.20 and out on the wild altiplano, the train breaks down. The Hitachi-built diesel had died and a replacement is summoned from the lineside phone, arriving 45 minutes later, but the passengers are now a nuisance. The train is split at Rio Mulatos and our new loco is taking off the goods waggons to Potosi, leaving us in the desert.

This outrage prompts a sit-down strike – in front of "our" engine – by the irate smugglers. There is a small riot in the office of the Jefe de Estacion. At 15.35, rocks are placed on the line to prevent the engine leaving, some sit on the tracks and a handful of women occupy the cab. The driver shuts off the power of his Sulzer 1985 Bo-Bo diesel. Amid the mayhem, we manage to get a fried egg sandwich apiece, officialdom finally relents at 17.08 and we set off again. The altiplano is changing, overcast and rainy with small lakes and streams. We are still over 11,000ft up, passing Rio Marquez, 3,798 metres, a hamlet of thatched houses, some derelict. Quillacas follows, evidently an abandoned mining town and at Challapata, past 20.00, five more policemen in army fatigues get on to check luggage again.

It's gone midnight before we reach Oruro, where we make a quick dash into the town to buy food, a risky business, and when we get back Coach 1401 is seething with locals. It stinks like a slaughterhouse as someone's contraband meat is rotting in the fetid air. I am obliged to evict a fellow who assumed, quite rightly, that we had disembarked, but he is quite decent about it and anyway another coach is added, easing the overcrowding. Another hour's shunting and we are off to La Paz at 01.40, and I go to sleep with a small towel round my face to keep the smell at bay.

When I awake at 06.00 on the outskirts of the Bolivian capital, there's yet more shunting and many of the locals leave the train. We have now been nearly 47 hours "on the road". Finally, we arrive at La Paz station at 09.12 local time. We take pictures on the platform, shake hands – and heads in disbelief that the journey could take so long. But we've done it!

And it seems that now, no one will, because the International has been discontinued on the Chilean side, with services in Bolivia from Oruro titled the Wara Wara del Sur.

CHAPTER 27

GREAT LITTLE TRAINS

IF THE PHILOSOPHERS are right, and small is beautiful, then Britain is a land blessed with much beauty, in the shape of the aforementioned narrow-gauge railways. Often trundling noisily through delightful scenery, they are the Gulliver's Travels of the iron road, the passion of a special kind of enthusiast, who tend their tiny engines with loving care.

Some, like the famous Great Little Trains of Wales, were first and foremost industrial enterprises, built to exploit the mineral wealth of the mountains, bringing enormous loads of slate down to the ports. Others were constructed as a cheaper, easier form of railway, like the Lynton and Barnstaple, the jewel of the north Devon coast, and the network on the Isle of Man. Occasionally, they are the whimsical brainchild of wealthy men, like the Romney, Hythe and Dymchurch on the shingle coast of Kent.

In restoration times, narrow gauge railways have been

opened on the abandoned trackbed of standard gauge lines, like the South Tynedale railway, high in the north Pennines on the former route from Haltwhistle to Alston. Some, like the Sittingbourne and Kemsley, have taken the place of standard gauge on the route to an industrial centre, in this case a Thames-side paper mill. Yet others, like the Great Whipsnade Railway, were installed as an extra attraction at a visitor centre, here the country outpost of London Zoo.

There is no end to the variety of gauges, settings or traction. That is the great appeal of the great little trains. And there are many of them – more than 50 at the last count (they sometimes come and go, amid mixed fortunes, like the Cleethorpes Light) – so you are never very far away from one. From the Lappa Valley, a "hidden gem" in rural Cornwall, to the Almond Valley in East Lothian – via, naturally, the Leighton Buzzard in Northamptonshire – there's one close by. Pretty well one for every week of the year, in fact!

As with standard gauge operations, we puffer nutters have our favourites. I'll take a look at a few of those on which I've travelled, sometimes quite recently, sometimes long ago, with children – or even as a child. Into the last category comes the North Bay Railway in the resort of Scarborough, east Yorkshire. This well-nigh perfect replica of the real thing, built by forward-thinking borough councillors in the recession-hit year of 1931, to a gauge of 20 inches, runs for just short of a mile from Peasholm Park to Scalby Mills right by the sandy shore of North Bay.

Oh joy, oh magic! To be taken on this thrilling ride as I was as a five-year-old (and often later, most recently as a

75-year-old, babe in arms), whooping with delight in the only tunnel, as the locomotive – green-liveried Neptune or Triton, the spitting image of Flying Scotsman – gives a toot on the whistle. Unforgettable. The line, which attracts as many as 200,000 visitors a year, was once used as a stage by Kylie Minogue when she appeared at the nearby Open Theatre. North Bay is now in private hands and the workshops are doing an excellent trade in supplying the narrow-gauge industry.

Fast forward to living in the south-east and the undoubted Queen of the Narrow Gauge is the Romney, Hythe and Dymchurch – the R&DHR, as it is known, quite like the grown-up thing. Constructed in 1924, to a 15-inch gauge, it runs for 13-and-a-half miles from the ancient Cinque port of Hythe to Dungeness, hard by the nuclear power station and lighthouse of that name.

This incomparable, hugely-popular railway, has a chequered – indeed, chequered flag – history. Originally the dream of two racing drivers of the Roaring Twenties, Captain Jack Howey and Count Louis Zborowski (who died before it could open, killed in an accident on the Monza circuit), the R&HDR opened in 1927, and has been going ever since. During World War Two it was taken over by the military who operated a unique armoured train.

It has more than a dozen locomotives, steam and diesel, some modelled on American locomotives and others on LNER types such as A1 and A3 Pacifics with names like Green Goddess, Southern Chief, Typhoon and Hurricane. For many years, the R&HDR could claim to be "the smallest public railway in the world" until the title was taken by the Wells and Walsingham Light Railway in Norfolk.

For decades it operated a school transport service and today attracts 150,000 visitors a year. It has also made cameo appearances on TV, including an episode of the *Inspector Lynley Mysteries* and children's programmes. There is nothing childish about the railway, however, as I discovered on my comfortable ride across the endless shingle some time in the 1970s when I was snapped leaning on Northern Chief.

Having been brought up on the magnificence of Mallard, big, clanking War Department and Stanier Eight-Freight locomotives, towering above trainspotters at the trackside, I cannot claim to be a great aficionado of the narrow gauge. But they have a unique place in railway history, not least in wartime where they supplied the World War One trenches. Nor are they all British-built. That doesn't stop them being of national interest, indeed, fascination. Between 1890 and 1908, the Ottoman Empire constructed an 800-mile line from Damascus to Medina, in present-day Saudi Arabia, to a 3ft5-and-a-quarter-inch gauge. The "Pilgrim's Railway" – or the Hedjaz – was ostensibly built to transport devout Muslims journeying to Mecca, but its strategic purpose was to supply Turkish troops and their garrisons. And that was why it was blown up, repeatedly and successfully, by World War One hero Colonel TE Lawrence and his band of guerilla Bedouin tribesmen.

If small is beautiful, how alluring might be the miniature version? Totally, argue passionate railway modellers, numbered in their thousands if sales of kit and magazines are any guide. Obviously, there are great advantages in creating your very own system in your attic or shed. It's unique, it's under cover, you're the station master, driver and signaller. The trains run to your schedule, as frequently as you wish. You can perpetuate your memories of steam days and long-forgotten rural rides. You are the controller, but you don't have to be fat. It is a child's dream come true, for men – it's usually men – who want to freeze-frame the railway they remember.

The history of modelling goes back almost to the origins of the steam locomotive. Push-along engines and carriages began to appear after the 1829 Rainhill trials. These were really just toys, like the wooden ones still sold today. In 1830s Germany, metal versions appeared, made by pouring molten brass or tin into a mould. A full train had hand-carved wooden fittings attached to metal bases. These "carpet railways" were delicate and had no moving parts.

American modellers claim the invention of the first true model, by one Matthias Baldwin, creator of the Baltimore Locomotive Works in the 1830s. His early version was taken up by several toymakers and by 1856, George Brown & Co of Connecticut is credited with creating the first known self-propelled American train. However, they were – and still are – known as "collectibles" rather than railways in miniature.

The first documented model proper was the Railway of the Prince Imperial, built in 1859 by Emperor Napoleon III for his three-year-old son, also (surprise, surprise!) Napoleon, in the grounds of the Chateau de Saint-Cloud in Paris. It was powered by clockwork and ran in a figure of eight, much to the delight of the little boy, and presumably his father, who "bought it for his son" – an excuse sheepishly trotted out by grown men ever since.

Mass production had to wait until 1891, when the German firm Marklin, which did a profitable business in dolls' house accessories, shifted to manufacturing tinplate trains. Early models were made from stamped metal and powered by wind-up or clockwork. The age of the model railway had begun and it has never looked back, despite the end of steam and the great locomotive from which the models derived.

The Golden Age of Modelling is regarded – like the railway itself – from the 1920s to the 1950s – with finely-crafted and realistic designs that brought the real thing into the living room. Different gauges appeared and permanent layouts for the permanent way, set in classic backgrounds of stations, signal-boxes, streets and countryside scenes.

The world's oldest working model is a system set up by the Lancashire and Yorkshire Railway to train signallers. Built at the company's Horwich works and dating back to 1912, it was in use until 1995 and now resides in the National Railway Museum. The world's largest is the Miniatur Wunderland in Hamburg, one of Germany's most popular tourist attractions. It occupies more than 100,000sq ft of floor space and boasts 1,230 digitally-controlled trains running on 54,000ft of track. There are sections for various foreign countries and talk of one for Britain, where DB – Deutsche Bahn – already operates the real thing.

There is a serious market for the hobby, with dedicated magazines and several UK manufacturers of both parts and complete locomotives. New electric models, based on loco-motives that the older generation will remember in steam and everyone can now see on heritage railways, appear regularly. They're not cheap. An LNER streamlined A4 Pacific offered by classic model maker would set you back £349.99, but you could celebrate Railway 200 by buying a replica of the Stockton & Darlington pioneer Locomotion Number 1 for £184.99 and very fine it looks too.

They don't make a lot of hot air about it, but there are some noted celebrity modellers, among whom rock legend Rod Stewart is perhaps the most vocal. In an interview with *Railway Modeller* ['the number one model railway magazine'] he showcased his breathtaking passion named "Grand Street and Three Rivers Railroad" inspired by the cityscape of 1940s USA. Construction of the 27 x 62ft layout with a 900ft main line on his country estate has taken him three decades – and it isn't yet finished. He spends up to five hours a day on the work and says, "When I walk into my workshop, it's like entering the gates of heaven. It's the greatest hobby in the world."

Another music legend, Frank Sinatra, might agree with him. He too had a private passion for tiny trains, housed in a special building for his layout. Neil Young is not only a hobbyist but partner in a US model manufacturing company, and Jools Holland boasts a layout from London to Berlin. Actors including Tom Hanks and the late Gary Coleman join the celebrity line-up, with even Walt Disney going before them. Showbiz mogul Pete Waterman attempted a portable model world record on *BBC Breakfast*.

Here, I must confess never having owned a model railway. I was once given a toy train set, but it didn't last five minutes. And while I admire the skill and dedication of modellers, I much prefer the real thing with which I grew up: the great, clanking freight locomotives, the diminutive "Jinty" shunters and the ubiquitous mixed-traffic LMS Black Fives, of which 18 still survive and give sterling service.

But it's clear that modelling doesn't just satisfy nostalgic yen. It can be therapeutic. As described in *Heritage Railway* magazine [Issue 332, June 2025], David Thompson, a 92-year-old enthusiast turned an unused room in the Sevenoaks, Kent, care home where he lives with his wife Olive, into an amazing model railway layout. His "quiet, personal" project over two years created one of the most popular and lively spaces, where fellow residents are invited to help shape a growing miniature town.

An accountant during his working life, David wanted to leave something behind. "It's more than just trains," he says. "It's a place where people connect. I love seeing others enjoying it. It brings back memories for many, and it's wonderful to share something that brings people together." And manager Esther Adams agrees, "David's railway room has added something special to the home. It's such a big part of our community now."

The love of trains in action, you might say. I certainly would, as the passion for the railway is rekindled in old age.

TOP 10 GREAT LITTLE TRAINS... *A WRITER'S CHOICE*

1. Ravenglass and Eskdale, known locally as "la'al Ratty" is a 15-inch gauge line running through incomparable Lake District scenery from the coastal village of Ravenglass to Dalegarth near Boot.

2. North Bay Railway, 20-inch gauge, rattles by the seaside for just under a mile in Scarborough from Peasholm Park to Scalby Mills.

3. Welsh Highland Railway, a 1ft 11½-inch gauge miracle of restoration, winds its way through wild scenery for 25 miles from Caernarfon to Porthmadog.

4. Brecon Mountain Railway, in South East Wales, deserves its regular appearances on TV. Running on an old BR trackbed, the 1ft 11¾ gauge line climbs from Pant, near Merthyr Tydfil to Torpantau.

5. Cleethorpes Coast Light Railway is a 15-inch line taking trippers from the Lincolnshire resort's Leisure Centre to Lakeside station where you can find the Signal Box Inn, claimed to be the smallest pub on the planet.

6. Volk's Electric Railway in the south coast resort of

Brighton is in a class of its own. The world's oldest operating electric system, this brainchild of Dr Magnus Volk built to a 2ft gauge in 1883, takes holidaymakers for a mile-long ride along the seafront. Don't get your feet wet!

7. Devon's Lynton and Barnstaple is a 1ft 11½-inch railway, constructed in 1898 by media magnate Sir George Newnes and closed by Southern in 1934, that refused to die. Enthusiasts have reopened a one-mile section from Woody Bay to Killington Lane and hope to rebuild the full 19-mile length through rolling countryside eventually.

8. Bure Valley Railway, England's second-longest, takes visitors nine miles on a 15-inch line from Wroxham to Aylsham on a long-closed BR trackbed. A gem in rural East Anglia.

9. Wells and Walsingham Light Railway, a 10¼-inch line, steams four miles from the coastal town of Wells-Next-The-Sea, in north Norfolk, to the beautiful Abbey village of Walsingham, for centuries a centre of Christian pilgrimage.

10. In Bedfordshire, the Leighton Buzzard Light Railway, a 2ft gauge line built in 1919 to serve sand quarries north of the town, is now run by volunteers who run trains on tracks just under three miles in length. John Travolta took his young son for a ride, and if it's good enough for the stars...

CHAPTER 28

THE GREAT ENGINE BUILDERS

IN OUR DISMAL times, celebrities are created from nobodies overnight by crude "reality" TV shows. In the Great Railway Age, the heroes were engineers who changed the world, mostly for the better. Men like George Stephenson, creator of the world's first public railway, Isambard Kingdom Brunel, who hurried people from London to Bristol on his 7ft gauge and spanned the Avon Gorge, and Sir Nigel Gresley, designer of the world's fastest steam locomotive.

They engineered a social revolution, bringing labourers from the fields into the towns and cities, they made the Industrial Revolution work, turning Britain into a great economic, military, and colonial power. Their vision of the future brought prosperity, and war, to far-flung places. Without them, there would have been no Empire, no global economy, for good or ill.

In the beginning of the steam railway, the driver was called the engineer. He still is in the USA, and the job description survives today over here in the name of the British drivers' trade union, the Associated Society of Locomotive Engineers and Firemen (ASLEF), even though there are no steam locos on the public railway. But by the end of the Victorian era, the most important man was titled the Chief Mechanical Engineer. He (it was always a "he") designed and built the machines that eventually ruled the iron road roost across the known world.

From being largely anonymous figures, working away in the obscurity of heavy industry, they were the celebrities, famous for their powerful, handsome locomotives, models of industrial art – and capable of setting world records for speed, hauling trains that offered the last word in contemporary comfort and leisure. Creators of the streamlined Pacifics dominating the East and West Coast Main lines were knighted and nameplates honoured Sir William Stanier of the LMS and the LNER's Sir Nigel Gresley on the engines they built. They were the rock stars of the steam railway. And while we have heard some parts of their stories, their lives bear further exploration.

This great lineage began long before the arrival of the world's first passenger railway, arguably in a corner of the Black Country in 1712, when ironmonger-inventor Thomas Newcomen built the first stationary steam engine, to pump water out of a coal mine near Dudley. It was so successful that hundreds more were installed in lead, coal and copper mines from Cornwall to Newcastle upon Tyne. Newcomen engines cornered the market until 1775, when James Watt, engineer son of a wealthy Clyde shipbuilder produced a more efficient version that superseded Newcomen's design. A "man o'parts" as they say in Scotland, Watt, an ingenious instrument maker at the University of Glasgow, also invented the concept of horsepower and the SI unit of power, the watt, is named after him.

Stationary engines became a universal requirement in the mining industry, creating in turn a new breed of keen experimenters with this tried and trusted form of power. Could steam be harnessed to a moving engine?

The principle of travel on rails was well established by the reign of George III. In July 1803, the Surrey Iron Railway opened a nine-mile goods line between Wandsworth and Croydon, now part of the south London metropolis but then largely open countryside by the river Wandle. But this was horse-drawn and could only travel at 2.5mph on flanged rails. Even so, it was a big improvement on canal or turnpike road transport.

Inventor Richard Trevithick, a champion wrestler born into the Cornish mining industry, has probably the best claim to the paternity of the railway. He first built a steam-powered road machine, the Puffing Devil, which carried six people up Fore Street, Camborne on Christmas Eve in 1801 and went on to make the first high-pressure engine before producing the first working steam locomotive. The unnamed engine hauled five waggons loaded with 10 tons of iron along almost 10 miles of the tramway of Pen-y-darren Ironworks in Merthyr Tydfil on 21 February 1804, thereby winning a 500 guinea bet for his ironmaster backer, Samuel

Homfray. The cast iron plates of the tramline broke under the weight of his invention and haulage returned to horses but a full-scale replica in the Welsh Industrial Museum, in Cardiff, is in steam occasionally.

Trevithick, a 6ft 4in giant, later pursued a remarkable career in central and south America, fighting in the army of Simon Bolivar and almost losing his life to an alligator. He died, broke, in the Bull Hotel, Dartford, Kent and lies in an unmarked grave there.

There was more to come. The pioneers were in a hurry. Matthew Murray, a machine tool manufacturer in Leeds, is credited with building the first commercially-viable steam locomotive, the twin-cylinder Salamanca of 1812, supplied to John Blenkinsops's Middleton colliery in the south of the city. It could pull many times its own weight, and worked for many years.

It wasn't called such in those days, but industrial espionage was rife in the burgeoning railway industry. They freely stole ideas from one another. A later Murray model strongly influenced George Stephenson.

The Stockton and Darlington then comes into the picture like a roaring lion. Timothy Hackworth, son of a Durham colliery blacksmith was appointed the first locomotive superintendent of the railway in May 1825, after previously building two colliery engines, Puffing Billy and Wylam Dilly (museum exhibits in London and Edinburgh respectively). Working at Robert Stephenson's Forth Street Works in Newcastle, he ensured the success of Locomotion Number 1 and three succeeding models on the S&DR. He went on to form his own business in Shildon, famously building the first locomotive for a railway in Russia in 1836 and exporting one of the first to Canada. They got about, these pioneers.

The names of George and Robert Stephenson, father and son, will forever be associated with engine design and development. Indeed, George is often described as The Father of the Railway. Born in Wylam, Northumberland in 1781, of parents who could not read or write, George was himself illiterate until the age of 18. He paid for his own education while working at the pit and the son of a colliery stoker taught himself to be a colliery engine-wright, admired for his knowledge and mechanical skill.

He invented a safety lamp to save lives underground and built the locomotive Blucher in the workshops of Killingworth colliery in 1813, but his chief contribution was a succession of crude steam locomotives to run on wheels, culminating in the building of an eight-mile track for Hetton colliery in 1822 that did away with horse-power. It was the first mechanised railway. His great triumph was the building of Locomotion Number 1 at his Newcastle works.

George went on to achieve even greater things: building the Liverpool and Manchester by floating the line across impassable Chat Moss. He bought a coalfield in Leicestershire, and revolutionised mining techniques. He even brought the York and North Midland Railway to Normanton, in what was to become the bottom of my street. He was the first president of the Institution of Mechanical Engineers and was lionised in a biography by Samuel Smiles, the Victorian advocate of self-help – of which there could have been no greater example than Stephenson – in 1857. He married three times and died, a rich man, aged 67, in 1848 and is buried at Holy Trinity church, Chesterfield, where he is remembered in this bicentenary year and by a statue in the National Railway Museum.

George's only son Robert followed in his father's footsteps. They were big shoes to fill, but he proved equally ambitious. He studied at Edinburgh University and in 1823 became managing partner of the family business in Newcastle, where Locomotion Number 1 and the passenger coach Experiment were built for the S&DR.

In the 1820s, he travelled to South America and the USA looking for business opportunities and on his return won the Rainhill locomotive trials with the Rocket he designed and built at his Newcastle works.

He went on to construct the Canterbury and Whitstable Railway, and still not 30, contracted to build the London and Birmingham Railway, which opened in 1844, having cost £4.4 million, more than twice the original estimate. Trains still pass through his tunnels and the notoriously-deep Tring cutting. By now famous, he travelled across Europe advising kings and companies how to bring their railway to their country. It wasn't all plain railroading, however. A bridge he built across the river Dee outside Chester gave way in May 1847. The driver and fireman survived, five passengers died in coaches that fell into the river.

Undaunted, in 1849 he constructed the wrought-iron Britannia bridge high across the Menai Straits for Irish Mail trains to Holyhead. It is still in use today, one of the wonders of the engineering world. His 1,372ft road and rail bridge 145ft above the Tyne into Newcastle of 1849 and 28-arch stone viaduct over the Tweed to Edinburgh were both opened by Queen Victoria, who graciously offered him a knighthood. He declined politely, as his father had done. But he did become the MP for Whitby, unopposed, for the rest of his life. Restless Robert advised against a Suez canal, arguing that it would fill with sand, but he did help build a railway from Cairo to Alexandria.

Returning from another triumph, the Norwegian Trunk Railway, he fell ill and died aged 55 on 12 October 1859, and was buried in Westminster Abbey, at a service attended by 3,000 admirers. He is remembered by a statue at Euston station.

The Stephenson legend is unrivalled, except by the extraordinary life and achievements of Isambard Kingdom Brunel, builder of the London to Bristol route in 1834, but he is more famous for his civil engineering triumphs like the two-mile Box Hill tunnel than locomotives.

Born in Portsmouth in 1806, the son of an English mother Elizabeth Kingdom and a French engineer father who worked on the first abortive Channel Tunnel, he was a child prodigy who absorbed the principles of engineering aged eight. After graduating from Caen university, he joined his father who was chief engineer for the Thames Tunnel between Rotherhithe and Wapping, working underground in hellish conditions. He narrowly escaped death in an accident in 1828 and while recuperating he made a design for the Clifton Suspension Bridge, opened in 1831. The magnificent bridge over the Avon and the tunnel are both still in daily use.

In railway circles, however, he is best remembered as the true begetter of the Great Western Railway, the only one of its kind to earn that soubriquet. Appointed chief engineer to the proposed London to Bristol line in 1833, he spurned standard gauge used everywhere else and set his rails 7ft ¼ ins apart, for greater speed and safety, plus increased goods capacity. Bristol was then the chief Atlantic port and Brunel's ambition was to sell through tickets from London to New York, on his train and his iron ship, the Great Western. His transatlantic vision never paid off and his broad gauge was eventually abandoned at great cost, long after his death, but his passengers to Bristol from the majestic trainshed he built at Paddington had the smoothest, fastest ride on the railway when it opened in 1844.

Arguably his finest achievement was that London terminus of the GWR. Ambitious, wilful Brunel announced, "I am going to design, in a great hurry and I believe to build a station after my own fancy, that is, with engineering roofs, at Paddington." It stands today. And only recently did TV's Jeremy Clarkson declare on Instagram, "By far the coolest thing in London. Brunel's roof at Paddington Station."

Not all his inventions proved successful. His "atmospheric railway" using stationary engines and pneumatic pressure to propel the trains from Exeter to Plymouth in Devon failed and cost the company half a million pounds. All that is left is a pumping station by the line at Starcross, but his tubular bridge across the Tamar into Cornwall in everyday use still bears the legend IK Brunel 1859. A giant metal signature to a giant of the railway, it is an honour to ride below it.

Celebrated by his peers, but not quite so much by GWR shareholders who bore the financial burden of his fertile brain, Brunel, a heavy smoker – famously with a cigar in his mouth – died aged 53.

One obituarist noted, "The history of invention records no instance of grand novelties so boldly imagined and so successfully carried out by the same individual." He is revered to this day, coming second, after Winston Churchill, as the second greatest Briton.

Streets, shopping centres are named after him. There are at least eight statues of him and the topmast of his revolutionary Great Eastern ship graces Liverpool's Anfield ground as a flagpole. He is remembered in a window in Westminster Abbey and his scientific vision is enshrined in the title of Brunel University, London, of which he would have been particularly proud. But his real tribute is the railway that carries millions every year.

By the mid-19th century, locomotive building had come of age. Superbly-designed engines that would last well into the 20th century were being constructed at the big operating companies' own works that had sprung up across the system: Glasgow, Manchester, Swindon, Bolton, Derby, Doncaster, Brighton, London and Crewe. A new class of locomotive superintendent had come to power: Daniel Gooch, a pupil of Robert Stephenson, at Swindon, Matthew Kirtley at Derby, succeeded by Samuel Johnson, Richard Deeley and Henry Fowler; Francis Webb, designer of the hugely successful 2-4-0 "Jumbo" class at Crewe; father and son Joseph and William Beattie at Nine Elms, for the London and South Western Railway.

At the turn of the century, the railway was hitting the newspaper headlines. The Great Race to the North of the 1890s didn't quite make the chief mechanical engineers into household names, but in the industry they commanded respect not far short of adulation. At Swindon, William Dean and his classic Goods and GJ Churchward built a solid foundation of standardised styles. On the companies of the south, Drummond, Urie and Marsh pitted their skills against competitors, as did Aspinall on the Lancashire and Yorkshire Railway, Manson on the Glasgow and South-Western, and Bowen-Cooke on the London and North Western Railway. The Quaker brothers Thomas and Wilson Worsdell ruled the roost at the North Eastern Railway before Sir Vincent Raven, while at the Great Eastern James Holden, creator of the successful Claud Hamilton class and his son Stephen, yet another Quaker held the reins for well-nigh three decades.

True celebrity only came to the unsung chief mechanical engineer after the Big Four companies were formed in 1923 and fierce competition ensued. This was the era of the two great names that reverberate a century later: Sir Nigel Gresley and Sir William Stanier. Gresley, born in Edinburgh in 1876, was a child of the Victorian age with a truly modernist vision. After an apprenticeship at Crewe, and various posts with the Lancashire and Yorkshire, he succeeded Henry Ivatt as chief mechanical engineers of the Great Northern in 1911, chiefly producing elegant "workhorse" goods and suburban traffic engines until his Pacific debut with the A1 in 1922, the forerunner of the Flying Scotsman class he created on becoming CME of the newly-created London and North Eastern a year later. Thereafter, he went from strength to strength, as did his locomotives, creating class after class of handsome passenger, freight and mixed traffic designs, culminating in the streamlined high-speed A4 Pacifics in 1935. These instantly-recognisable engines, with corridor tenders allowing crew changes without stopping, revolutionised travel on the London-Edinburgh East Coast Main Line, with non-stop services between the two capitals. Gresley was fond of breeding wild ducks on his Hertford-shire estate and one of his A4s, 4468 Mallard, set the world speed record for steam in 1938, which remains unbeaten. Knighted in 1936, he is commemorated by a statue at Kings Cross station (missing the original mallard at his feet that Gresley's grandsons found demeaning to his memory) and a plaque at Edinburgh Waverley station that pays tribute to a "an inspiration to generations of engineers who admire fine engineering and beauty of line".

Gresley's great rival, William Stanier, was born in the same year, in the railway town of Swindon, where he followed his father into the GWR works, rising to become manager in 1920. In a move characteristic of railway management, he was poached by the London, Midland and Scottish to become CME in 1932, charged with modernising the company fleet on Swindon lines. This he did, with a string of successful standardised freight and passenger locomotives, including the Royal Scot and Jubilee classes, the ubiquitous mixed-traffic "Black Fives" of which no fewer than 842 were built – 18 surviving on heritage lines today and mostly operational – and the 852-strong "Eight Freight" heavy goods class, sometimes called the 'Engines that Won the War' which also saw service overseas during World War Two.

But it was his Princess Royal Pacific class of 1932 and the even more powerful, streamlined Coronation class of 1937, built to run non-stop between Euston and Glasgow for which he became most celebrated. Determined to outdo Gresley, his 6220 Coronation Scot captured the speed record from the LNER with a 114mph run in 1937, but he only held it for a year before Gresley took it back. Records aside, Stanier – knighted in 1943 – is most admired for bringing Swindon-style standardisation of design and construction to the LMS, and subsequently to the nationalised British Railways, where CME Robert Riddles continued his methods right up to the end of steam at Swindon in 1960.

No list of great engineers can be complete, however, without a mention of Oliver Bulleid, the brilliant, eccentric CME of the Southern Railway. Born in 1882, he was another graduate of the Doncaster School, becoming manager there in the early 1900s. After Great War service in the army transport arm, a spell at GNR and various jobs in Europe, he was brought back to 'The Plant' by Gresley as his assistant in 1923, working on the pair's most successful designs. He was poached by Southern Railway in 1937, introducing his controversial Merchant Pacifics, with air-smoothed streamlining and many novel features learned from his European experience, a year later. These striking locomotives, nicknamed "spam cans" by enthusiasts, were followed by the smaller West Country 4-6-2s. They were eventually rebuilt on more orthodox lines and many survive on heritage lines. His equally-arresting Q1 class 0-6-0 was perhaps the strangest-looking engine to emerge from a British manufacturer.

Bulleid took his far-sighted strategy further than steam, accelerating Southern's electrification programme and developing diesel-electric traction. However, his double-ended Leader locomotive, built in 1948 and looking more like a modern diesel but powered by steam, proved an innovation too far. After "difficult" trials, it was scrapped. So, too, was his turf-burning Irish version built during his time as CME of the Irish Railways in the 1950s. Bulleid died in retirement in Malta, aged 87, in 1970. In his obituary in *The Times*, he was described as "the last truly original and progressive mechanical engineer of the steam locomotive era in Britain".

Remarkably, steam engine production continued on British Railways long after it was abandoned elsewhere, under CME Riddles, another graduate of the Crewe school. During World War Two, as director of equipment at the Ministry of Supply, he had introduced the Austerity fleet of heavy goods engines, and convinced chiefs of the newly-nationalised railway that steam was cheaper and simpler than diesel and electric traction.

BR rationalised its repair and manufacturing arm, but coal-fired locomotives continued to pour out of Derby, Swindon, Doncaster, Crewe, Darlington and other works including Brighton until 1960. In all, 773 'Standards' were built ranging from humble two-mixed traffic tanks through the copy-cat Black Fives to the mighty Britannia Class Pacifics and the one-off Class 8 Duke of Gloucester, out-shopped at Crewe in 1954. Though largely successful, most of this huge cohort lived short lives, sometimes of only five or six years, but 46 survived into preservation, where they continue to give excellent service.

And the Riddles design endures, in the shape of a £5.5million new-build Clan Class Pacific, 72010 Hengist, under construction at a specialist manufacturer in Sheffield and scheduled for steaming in 2030. Construction of a new 2MT tank 84030 is also nearing completion and many other new-builds include an A1 Pacific Tornado and Gresley-designed Sandringham, a Highlander, a P2 2-8-2 of the LNER, Betton Grange and County of Glamorgan of GWR, and Patriot class The Unknown Warrior of the LMS.

The age of the Chief Mechanical Engineer is not dead. It lives on, on the footplate of almost 400 British locomotives saved from the cutter's torch for our national heritage, plus more than 30 new-builds taking shape from the drawing board of yesteryear. Not just museum pieces, but steaming, smoking, clanking reminders of the Great Railway Age ushered in by a contraption built in a shed 200 years ago for an obscure colliery line in Durham.

Between them, these men of the Railway Age did more to shape the future of their country than monarchs, prime ministers and bishops, with their faith in science, engineering, business acumen and sheer nerve.

CHAPTER 29

GO FASTER! THE FUTURE OF RAIL

SPEED IS WHAT it was always about: how to get produce to market and people to their destination as quickly as possible. That way lay business success. The age of the horse and cart, or carriage, had lasted for millennia, ever since ancient man invented the wheel, and its time was up. With the world's first public railway, Britain showed that commerce and travel could be accelerated beyond our wildest dreams. And in the century that followed, our railway set world speed records – for steam.

But then the steam went out of it. While other countries, notably Japan and France, went full-tilt for high-speed trains in post-war years, it took half a century before Britain got one and 30 years after that the second one is still being built amid endless delays and acrimonious controversy.

HS2, the line that was going to revolutionise rail travel between London, Birmingham, Manchester, Liverpool, the East Midlands, Leeds and York, and potentially Scotland, is in a tunnel somewhere in the Chilterns, with trains not due to arrive at restored Curzon St station in "Brum" until 2033 at the earliest. Then there was HS3, which would supposedly connect Manchester and Leeds and possibly Hull, which turned out to be not pie in the sky but pie in the Pennines.

What happened? British Rail did get the highly-successful High Speed trains, capable of 125mph in the 1970s and still operational on some cross-country routes. The East Coast Main Line from London to York, Leeds, Newcastle and Edinburgh was progressively electrified in the 1980s with a new fleet of electric trains capable of 140mph, but on safety grounds limited to 125mph.

State rail chiefs also commissioned the Advanced Passenger Train, designed to run at 155mph. But the APT was beset with technical and financial problems. VIPs on the press trip complained of feeling sick when it tilted and amid negative media coverage the project was abandoned.

It was not until the end of the century that Britain's first truly high-speed line, HS1, was built, and then partly out of national embarrassment. The Channel Tunnel, completed in 1994, was served by high-speed trains on the French side, but slow-coach this side. Construction began from the Kent coast in 1998, reaching London in 2007 to a vastly restored and redesigned St Pancras at a cost of £5billion. It was an instant success, with Eurostar taking passengers to Paris in two hours and 15 minutes.

HS1 was only 62 miles long and so obviously required to forestall Gallic satire at the expense of les Anglaise's that it attracted only moderate environmental and political opposition. The same was not true when HS2 was mooted. The tree-huggers, bat and newt protectors and MPs in seats "threatened" by the line turned out in force to frustrate, delay and if possible halt the project.

But their objections were as nothing compared to vast scale of political, financial project mismanagement that made HS2 the laughing stock of the railway world. Construction did not begin until 2020 and costs rose dizzyingly from £20.5billion in 2012 just for Phase 1, to between £49 and £57billion a decade later, and then by another £10billion in 2024. Stage by stage, arm by arm, the project was dismembered. Links to York, Leeds and the East Midlands, Manchester and Liverpool, and finally Crewe, ancestral home of the LMS which HS2 was proposed to complement if not replace, were abandoned.

Under the attenuated scheme, HS2 Ltd aimed to have trains running into Birmingham by 2030, or perhaps three years later. And not from Euston, as originally planned, but from Old Oak Common in west London, on the site of the GWR's largest and oldest steam engine shed – inconvenient for travellers and very far from what the nation was promised, demonstrating yet again that politicians and railways are a toxic mix.

Coming into power in 2024, Labour ordered yet another review by a new chairman of HS2 Ltd, the government-owned construction company, Mark Wild. A year later, he produced the most damning report imaginable. The project was "an appalling mess", running wildly over budget. Trains would not run in 2033 and no firm date could be set for when they might. They would travel at reduced speed – 200mph, not the promised 248mph. The pared-back scheme offering only London to Birmingham was now estimated to cost at least £100billion.

Billionaire Michael Goss, who owned the Euston station site before it was compulsorily purchased for HS2, commented, "The government knows that HS2 was a Ponzi scheme from the beginning. It's a gravy train to nowhere and it will never happen. The country can't afford it." Amid allegations of civil servants and contractors lying and covering-up, Prime Minister Starmer ordered a top-level inquiry by the Cabinet Secretary to examine if civil servants and other public bodies should face investigation into their roles in the chaos. To date, no one has been punished for this fiasco-on-wheels (and there are none of those either) – except for a few whistleblowers who exposed the truth.

It gets worse. HS3, never more than a pipe-dream, was ditched in favour of TPU – the TransPennine Upgrade, electrifying the 76-mile route from Liverpool through Manchester and on to Leeds, installing digital signalling and improving stations. The first phase costing £12billion is under way and before the general election, ministers approved a scheme to electrify the line east from Cotton-opolis to the Pennine summit, where trains would switch to diesel. The promised year of disruption are on schedule.

Renationalisation of the railway, or de-privatisation, depending on your point of view, as promised by Sir Keir Starmer in Labour's 2024 election manifesto, will almost certainly take longer and cost more than predicted. Everything to do with the railway always does.

Labour's pledge does not amount to total renationalisation on the scale of 1948. Freight, a very profitable business, will stay in private hands, so will the dividend-hungry train leasing companies. Essentially, it is the passenger business and the infrastructure being brought under one roof, with Great British Rail [GBR] having overall authority. There is still large scope for confusion about who does what and where the subsidy goes. Political purists might say this is pseudo-nationalisation because it doesn't go the whole hog, but it's the best on offer.

Ministers have given "Scarborough warning" to rail unions that the coming of GBR, based in Derby after a competition for choice of HQ, isn't all good news. Recreating a form of British Rail owning all the passenger franchises as they come up for renewal will end duplication of teams and mean "multiple thousands" of redundancies. There will also be job losses at HS2, a "government source" told *The Times*, adding that they will be mostly in back-office roles where staff are not members of the most active trade unions – meaning Aslef and the RMT. Some rail quangos, including Transport Focus and the Rail Ombudsman (who knew there was one?) will be absorbed in a new Passenger Standards Authority. This indigestible mass of measures features in a Railways Bill "this year", at a time when the new GBR logo is applied. Branding is everything these days.

Perhaps because it has been with us so long and been so intertwined with our lives, the railway has become part of the family. Line closures have been treated like a bereavement, with crowds turning up for the last train, decorating the locomotive and even carrying a fake coffin to signify their social grief. By contrast, there is no record of people marching behind a banner demanding "Take Away Our Railway" or campaigning to shut down their local station, with letters to their MP and the newspapers insisting that signallers and porters should be sacked.

In fact, when the rail bosses, anxious to please their political masters with large-scale cost savings, the public revolts, and not just the travelling public. In the autumn of 2023, the Rail Delivery Group proposed the closure of almost all the network's ticket offices – 974 of the 1,007 in total, with the loss of thousands of jobs.

A public outcry followed, with more than 750,000 people taking part in a consultation that was meant as a blind, softening up opinion, with 99% expressing hostility. The ticket clerk was for them the human face of the railway, real people like my father Harry who was a ticket clerk for many years at stations in West Yorkshire. Bland, reassuring talk of substitute "welcome points" of staff on the platforms cut no ice. That job had been done for generations by porters – and they had all been done away with.

For once, passenger watchdogs Transport Focus and London TravelWatch bared their teeth and the government was forced into an embarrassing U-turn, ordering the train operators to withdraw the plan. Rail bosses were furious, claiming that the Department of Transport civil servants had signed off the shutdown, in order to save on subsidies.

The picture is not as dismal as the cruel farce of HS2 suggests. The tide of station closures has been halted and reversed. More are reopening, to huge social and economic effect. My own local station, Cononley in North Yorkshire, shut down in the Beeching era, reopened in 1988. The village has experienced a building boom, with a disused mill across the road from the platform to Leeds converted into award-winning apartments, with a small estate behind it. Next stop up the line, Steeton and Silsden, also closed and reinstated, now has a multi-storey car park for commuters. Historic Keighley has just had a £10.5million facelift.

The future's bright, the future's orange – everywhere you go by train these days you see orange hi-viz jackets and construction going on. New lines are opened, or promised. The last train to Exeter left Okehampton in 1972, but with funds from the Restoring Your Railway Fund, the Dartmoor Line resumed services in 2021.

The Northumberland Line linking Ashington, once the largest colliery village in England, to Newcastle, has been reinstated, attracting five times the number of projected customers. Most ambitious of all, reinstatement of the "Varsity" cross-country line from Oxford to Cambridge is under way. Disappointingly, Chancellor Rachel Reeves shut down the restoration fund in her July 2024 budget, to save £85million public spending.

Renationalisation actually began on 25 May 2025, with the media eager to pounce. At 01.59 South Western Railway became the first to be taken back into public ownership, but the first train – the 02.27 from Guildford to London

Waterloo was cancelled and replaced by a bus due to "maintenance and upgrades". There was irony at work here. When the first train under privatisation arrived at 6.00 at Waterloo in 1995, Tory Transport Secretary Sir George Young greeted the first passenger for the media – only to discover he was a tax dodger!

Renationalisation and commemoration came together on New Year's Day 2025, with the kick-off for Railway 200, the ambitious project to celebrate the bicentenary of the world's first public railway. At midnight on 1 January 1948, drivers and firemen blew engine whistles all over the country, a cacophony of joy over the advent of nationalisation of the railway. On 1 January this year, The Great Whistle-Up replayed that festive cacophony of joy. The noise of rejoicing rang out at noon, sounded on more than 200 steam, diesel and electric preserved locomotives at 60 venues, watched by crowds who braved the wind and rain to add their cheers to the din. In Brighton, television newsreader Nicholas Owen blew a ceremonial whistle at the Volk's Electric Railway.

An even more remarkable "double" commemorates the actual date of the railway revolution in late September 2025, when a working replica of Locomotion No 1 re-enacts the events of two centuries ago. This unique rebuild, constructed in 1975 for the 150th anniversary of the S&D, hauls a replica carriage, Experiment and three chaldrons – coal waggons – from Shildon via Darlington to Stockton on Tees, on Network Rail metals following the original route.

As a taster for the big spectacle, a Railway 200 train, titled Inspiration, made up of three specially decorated coaches tours the country until the summer of 2026, visiting 60 or more venues to a likely audience of 200,000 school children and members of the public.

Bicentenary Year jollifications like The Greatest Gathering in August would have astounded the strait-laced Quaker pioneers of Darlington. This three-day festival in Derby, featured locos plus industry chiefs, politicians and ace puffer nutter Pete Waterman, fairground attractions, street food and live music. The railway family knows how to party!

From the works of JMW Turner and Monet to the brilliant poster-art of Terence Cuneo, the railway has always been a fascination to painters. Railway 200 held a competition for the public's favourite UK railway-themed artwork. And the winner was: *Train Landscape*, painted by Eric Ravilious in the 1940s, depicting the famous Westbury White Horse through a third-class carriage window.

Heritage Minister Baroness Twycross hailed the artwork as a worthy victor, saying, "This evocative watercolour invites us all to experience a railway journey through an artist's eye, capturing a uniquely British perspective that resonates today. Art offers a powerful way to engage with our past, telling the unique story of Britain's relationship with the railway over 200 years."

My favourite commemorative artwork may be seen in my home town, Normanton, where artist Harriet Colours has painted a composite scene of locomotives, colliery works, coaling and ash plants and the magnificent station (now gone) on an entire wall of the refurbished market.

As if the public carnival were not enough, Railway 200 is

being commemorated on TV by, yes, you've guessed it, a new programme later this year presented by the Greatest Telly Rail Journeyer himself, former Tory MP Michael Portillo.

Every big event like the bicentenary brings a shedload of commemorative items, from mugs to tea towels. And you can even jingle the joy in your pocket or purse with a £2 coin struck by the Royal Mint depicting the big day in 1825 with the edge inscription Active – Locomotion No 1.

Main line or heritage, historic or modern, this is still the age of the train. For many of us, who would not dream of travel by any other means – except bus to the station – it has always been the age of the train. Always has been and always will be. They are our trains, and we love them.

An extra treat for fellow puffer nutters...

WEIRD AND WONDERFUL
THE GREAT RAILWAY QUIZ

QUESTIONS:

1. What is Railway Time?
2. What was the last steam engine built in Britain?
3. What is the UK's largest and busiest station?
4. Which politician first died in a railway accident?
5. Which bridge is most famously commemorated in bad poetry?
6. What was the Turbomotive?
7. What is Britain's remotest station?
8. Who was Big Bertha?
9. No one went and no-one came: where?
10. What was a "cop"?
11. What is Rain, Steam and Speed?
12. Who was Concrete Bob?
13. What is a Jinty?
14. Who were the navigators?
15. What is the deepest London Underground station?
16. Who celebrated foxhunting?
17. What do Copley Hill, Newton Heath and Sheep Pasture have in common?
18. Where could you go out to sea on a train?

19. What happened on the 4.50 from Paddington?
20. What is the difference between a saddle and a pannier tank?
21. How many people travel by train in the UK?
22. Who said "the only way to catch a train is to miss the one before it"?
23. What is the largest town without a station?
24. What is Britain's longest tunnel?
25. Where would you find the Crewe of the Coalfields?
26. Who or what is a gricer?
27. What is the most common name for a locomotive?
28. Which railway station is rising from the ruins?
29. Where was the rail transport driven by sail?
30. What is the least-used Network Rail station?
31. Who introduced sleeping cars?
32. What were the Flying Bananas?
33. What were the Flying Pigs
34. What were water troughs?
35. How many level crossings are there in Britain?
36. What was Motorail?
37. Why is scrap merchant Dai Woodham revered?
38. What outlandish invention did OVS Bulleid build for Irish railways?
39. Where were deceased passengers welcome?
40. Who died at Astapovo station, surrounded by media?
41. What was the shortest locomotive name?
42. Which station never saw any trains?
43. What were the Black Fives?
44. Which station makes passengers most happy?

45. Whose ashes were laid to rest in a locomotive firebox?
46. Who introduced the dining car?
47. What was known as Blue Streak?
48. What is the speed limit in the Channel Tunnel?
49. Where will I find a line from our salad days?
50. Who came second to Winston Churchill in a BBC poll of greatest Brits?
51. What was the target of the Golden Arrow?
52. What is a double-header?
53. What did George Bradshaw do for the railway?
54. How many bridges on Network Rail
55. Who installed the first Ladies' Waiting Rooms
56. What is the highest point on British railway?
57. What was the highest, until 1939?
58. What were the Crabs?
59. What were the Parliamentary trains?
60. What identity did Hogwart Castle steal?
61. What the Deltic?
62. Where is the original Metroland?
63. Who was Ruswarp?
64. Who published the Railway Review?
65. Where can you wait for a train and watch the barges sail below?
66. Why is Tornado not an ill wind?
67. Where is Britain's longest continuous platform?
68. What is a fireman's breakfast?
69. What connected dogs, bikes and prams?
70. What is a grampus?
71. Which locomotive starred in three films?

72. Which TV chatshow star was born on a London station?
73. What's a bull head or a flat bottom and sits in a chair?
74. Which famous railway photographer died on Settle station?
75. When was the first railway excursion?
76. Who belonged to the Railway Sleepers' Club?
77. What was called the unluckiest locomotive?
78. What proportion of UK train journeys start or end in London?
79. Where could you drink in the Fastest Bar in the West?
80. 80. What was "The Slow and Dirty?"
81. Which famous Victorian sibling was a booking clerk at Sowerby Bridge in the 1840s?
82. Which railway called itself The Premier Line?
83. Who said: "I see no reason to suppose that these machines will ever force themselves into general use."
84. Where did the film-famous Brief Encounter take place?
85. How did the footplate crew change over on the non-stop Kings Cross to Edinburgh services?
86. What was a railway roundhouse?
87. Who said "Even a journey on the Eastern Counties must have an end at last?
88. Which is the only station named after a work of fiction?
89. What were the Jellicoe trains?
90. What happened to the Royal Mail coach involved in the 1963 Great Train Robbery?

91. What is a ferroequinologist?
92. Which train was affectionately titled after a children's seaside entertainment?
93. Who said "Verily, railways are Abominations!"
94. What was the Atmospheric Railway?
95. When did the railway companies get above themselves?
96. Where do you find Poet's Corner on Network Rail?
97. What was the passimeter?
98. What is a diamond crossing?
99. Travellers south at Wakefield Westgate to London go over the famous Arches. How many arches are there?

ANSWERS

1. Synchronised standard time across the country, introduced by the GWR in 1840.
2. 92220, Evening Star, a 2-10-0 freight locomotive, made at Swindon workers and now in the NRM.
3. Waterloo, 24 platforms and 90 million users annually.
4. William Huskisson, MP for Liverpool, killed in an accident during the Rainhill Trials in 1830.
5. The Tay Bridge, lost in a storm in 1879, immortalised by Scots poet William McGonagall.
6. A Princess Royal Class which used steam turbines instead of cylinders. It was a failed experiment converted to normal traction as 46202 Princess Anne.
7. Corrour, on the west Highland line, a request stop with no road access.

8. A giant 2-10-0 LMS engine, reputedly the most powerful, built to bank trains up the Lickey Incline near Birmingham.
9. The station of Adlestrop, in the poem by Edward Thomas.
10. The trainspotter's word for an engine seen for the first time.
11. A famous oil painting of 1844 by JMW Turner, now in the National Gallery.
12. Sir Robert McAlpine, pioneer builder in concrete of the Glenfinnan viaduct in 1898.
13. An 0-6-0 tank shunting engine of the Midland Railway.
14. Usually shortened to "navvies" they were the 19th century skilled labourers who built the railways
15. Hampstead, 192 feet below ground level.
16. Sir Nigel Gresley, with his "Hunt" class locomotives of 1934.
17. They were all steam engine sheds.
18. Southend Pier, which has a 1.3 mile railway, the longest such in the world.
19. Murder, in Agatha Christie's 1957 "Miss Marple" novel.
20. Like horses, one has water slung over the saddle, the other hanging by its sides.
21. 3.5 million a day, the most in Europe.
22. The writer GK Chesterton.
23. Gosport, Hants.
24. The Severn Tunnel, between Bristol and South Wales,

4.5 miles long.

25. In Normanton, West Yorkshire. But not any more!

26. A trainspotter, who bags engine numbers like shooters and grouse.

27. Hercules, after the Greek god of strength.

28. Curzon Street, Birmingham, for HS2.

29. The light railway to Spurn Point, East Yorkshire.

30. Denton, near Stalybridge, with one train a week.

31. The Great Northern Railway, London to Glasgow in 1873.

32. GWR cream-and-brown streamlined railcars, introduced 1933. Three are preserved.

33. LMS 2-6-0 mixed traffic locos with novel raised running plates.

34. Water-filled gutters slung between tracks for steam engines to replenish on the move.

35. 9,000.

36. A car-carrying service, mostly to Scotland, made redundant by the motorways in the 1980s.

37. He saved from extinction more than 200 steam locomotives in his South Wales scrapyard.

38. An experimental turf-burning locomotive: it never entered service.

39. The London Necropolis Railway between Waterloo and Brookwood cemetery, 1854 to World War Two.

40. Russian novelist Leo Tolstoy, fleeing his wife, in 1910.

41. GNU, carried by LNER B1 locomotive 61018.

42. Dartmouth, Devon, for a line never built.

43. Stanier's most successful design for the LMS, a mixed-

traffic 4-6-0 of 1934, of which 842 were eventually built.

44. Kings Cross, with 96% satisfaction in a survey.
45. Alan Pegler, one-time owner of 60103 Scotsman.
46. The GNR, London to Leeds in 1879.
47. A Blue Pullman, running from London to Manchester and Bristol in the early seventies.
48. 99 mph.
49. The Watercress Line, a heritage railway in Hampshire.
50. Railway genius Isambard Kingdom Brunel
51. Paris, on the Pullman service from Waterloo, 1929-72, with a break for the war years.
52. Two linked locomotives pulling an unusually heavy train, of goods or passengers.
53. He introduced the first timetable in 1839 – and inspired the Portillo TV series.
54. 40.000, at the last count.
55. The Liverpool and Manchester in 1831.
56. Druimuachdar Summit between Perth and Inverness, 1,484ft above sea level.
57. Wanlockhead, 1,298 ft.
58. LMS 2-6-0 locos with inclined pistons that gave them a crab-like appearance.
59. Compulsory low-fare services imposed by Parliament in 1844, hated by railway companies.
60. That of GWR Hall 4972, Olton Hall, now exhibited at the NRM, Shildon.
61. A 3,300hp English Electric blue and white diesel the forerunner of a class that ousted steam from the East

Coast Main Line.

62. Commuterland outside London, a name coined by the Metropolitan Railway in 1915.

63. The faithful dog of Settle-Carlisle line supporter Graham Nuttall, who stayed by the body of his master for 11 winter weeks on a Welsh mountain, commemorated by a statue on Garsdale station.

64. It was the weekly newspaper of the National Union of Railwaymen distributed to members, certainly until the 1960s. My father took it.

65. Blackfriars station in London, on platforms built over the Thames in 1987.

66. In 2008 Pacific 60163 Tornado became the first steam locomotive built in England since 1960, in Darlington to the design of LNER engineer AH Peppercorn.

67. Gloucester, 1,975 ft [602 metres].

68. Egg and bacon cooked on a shovel in the locomotive firebox.

69. They were all Edmondson tickets under British Rail.

70. A railway waggon for ballast and sleeper

71. Lion, built in 1838 for the Liverpool and Manchester, appeared in Victoria the Great (1937), The Lady With The Lamp (1952) and The Titfield Thunderbolt (1952). Now on display in the Museum of Liverpool.

72. Jerry Springer, on 13 February 1944, in Highgate Underground a wartime refuge from German bombing raids.

73. A line on the permanent way

74. Bishop Eric Treacy, on 13 May 1978, of a heart attack

while waiting for a steam special. Restored Black 5 number 45428 is named after him.

75. On 14 June 1836, from Wadebridge to Wenford Bridge, Cornwall, when 800 cheap-fare passengers travelled behind locomotives Elephant and Camel.

76. Fleet Street newsmen who slept beyond their home commuter station after a night – or even a day – on the tiles. Longest miskip: a Reading resident who woke up in Swansea.

77. Diesel class 40 number 40126 headed the Great Train Robbery night mail, and was earlier involved in a crash near Crewe that cost 18 lives. It also ran away approaching Birmingham New Street.

78. 70%

79. The late evening trains from Paddington to Swindon with a refreshment counter.

80. The Somerset and Dorset Joint, running from Bath to Bournemouth through the Mendips.

81. Branwell Bronte.

82. The London and North Western.

83. The Duke of Wellington.

84. The refreshment room of Carnforth Junction, Lancashire, and still serving tea.

85. Through a corridor in the tender linked to the train.

86. A Victorian engine shed built round a turntable. Three still exist, one at Barrow Hill, Derbyshire, as a museum.

87. William Makepeace Thackeray

88. Edinburgh Waverley, celebrating Walter Scott's novel.

89. Secret specials of World War One, chiefly of coal but also crew, from England and South Wales to northern Scotland to supply the Royal Navy at Scapa Flow.

90. It was scrapped and burned in secret to avoid glamourisation of the £2.6million heist.

91. A cod-Latin term for a trainspotter.

92. The Marlow Donkey, a GWR branch line. The line is still open, but the 0-4-2 tank departed long ago.

93. William Morris, 19th century socialist and artist.

94. Brunel's short-lived pneumatic pipe traction experiment tried in South Devon in the 1840s. A pumping station survives at Starcross.

95. Faced with growing competition from airlines, the Big Four started regular air services in the 1930s. They were take over by BEA in 1947.

96. Statues of Poet Laureate John Betjeman at St Pancras, and Philip Larkin at Paragon station, Hull.

97. A booking office at some Yorkshire stations.

98. Where lines cross at right angles on the permanent way.

99. There are 95, verified by a local observer.

ACKNOWLEDGEMENTS

With thanks to Clare Fitzsimons, my commissioning editor at Mirror Books, text editor Nick Webster, and all my railway friends past and present, particularly the train drivers' union, Aslef. Also, it should be said, to the long-defunct Wakefield Railfans' Club, through which I met my wife of 62 years, Lynne, in the romantic era of steam.